PLANNING PRIMARY
SCIENCE

PLANNING PRIMARY
SCIENCE

Key Stages
1&2

Roy Richardson
Education Inspector (Primary)
Lincolnshire County Council

Phillip Coote
Deputy Headteacher
William Stukeley Primary School, Holbeach, Lincolnshire

Alan Wood
Advisory Teacher for Curriculum Management (Primary)
Lincolnshire County Council

JOHN MURRAY

Cover photograph: Manfred Kage, Science Photo Library. A leaf from the Kangaroo Vine showing electromagnetic discharge around its edge

Line drawings: Art Construction

Photograph (*p. 105*): Griffin and George

© Roy Richardson, Phillip Coote, Alan Wood 1993

First published in 1993
by John Murray (Publishers) Ltd
50 Albemarle Street, London W1X 4BD

Layout by Mick McCarthy
Typeset by Litho Link Ltd, Welshpool, Powys, Wales
Printed by St Edmundsbury Press
Bound by Hunter & Francis Ltd, Edinburgh

A catalogue entry for this title can be obtained from the British Library

ISBN 0 7195 5232 X

Other titles in the **Key Strategies** series

Primary Science A Complete Reference Guide by Michael Evans
Physical Education A Practical Guide by Elizabeth Robertson

Contents

Introduction

PLANNING FOR SCIENCE THROUGH KEY STAGES 1 AND 2

This support material is aimed to help primary schools review, or make a fresh start in implementing, the statutory orders for science. It sets out clearly the process which all schools will need to complete in order to write their science policy and scheme of work; both of which are legal responsibilities for all headteachers and governing bodies.

There are now legal requirements for all primary schools to give each child access to the Programmes of Study for science. Primary teachers need knowledge, confidence and guidance to give children this entitlement and to ensure quality of learning. We hope that the material in this book will assist teachers to plan and deliver this entitlement.

Schools have now received from the DFE folders for all the core and foundation subjects, each one demanding time and effort to implement. Many schools are therefore now taking a new look at each curriculum area and perhaps reviewing their original plans and styles of delivery.

It can be a lengthy task to map out the entitlement and headteachers, as curriculum managers, need to ensure that no sections or strands are neglected or omitted. Staff need the support of a long-term plan, or Programme for Science, which will save them considerable time in the early stages, and will provide a basis for planning scientific investigations and activities for all children.

To meet this need section 4 shows clearly how teachers can draw up a Programme for Science in which they can map out the entire content of science throughout Key Stages 1 and 2, irrespective of the size of the school or how children's age groups are arranged within it.

Time is a commodity which primary schools find is in short supply at present. An example Science Policy (pages 13–15) and Programme for Science (page 19) have been included which schools may choose to adopt, or adapt, to suit their own situation and ethos.

The example policy and Programme for Science have been used to create a Scheme of Work for Science (pages 25–136) which could again be used by schools of any size and any combination of mixed-age classes.

Section 5 gives examples of scientific activities and investigations that ensure coverage of all the Programmes of Study for science. The examples are set down in units of work matched to the Programme for Science so that teachers can see clearly what has to be covered and how. Also matched to each unit of work are the Statements of Attainment that can be assessed at each stage. The activities and investigations are designed to provide a minimum that schools will need to achieve in order to deliver their statutory requirements. Each unit can be elaborated if required but, with time at a premium on the primary school timetable, it is important that science is clearly planned and delivered to ensure coverage of all the Programmes of Study and that assessments are made against the Statements of Attainment.

Curriculum development is progressive and schools will need to set down a realistic timetable from its inception. Headteachers will need to ensure that science planning is given priority during this period.

We hope that this book will assist primary school teachers to draw up or review their plans for science so that ultimately children will benefit from an improved quality of teaching.

Roy Richardson, Phillip Coote and Alan Wood

Before making a start...

The role of the headteachers in any curriculum development is crucial and without their total commitment, support and guidance it is pointless making a start. Headteachers need to give their time to define the timescale of teaching and ensure that priority will be given to achieving the required standards.

Drawing up a science policy, or dividing the science statutory orders into units to be taught by individual teachers, is relatively easy and takes little time. The pressure on primary schools to implement all areas of the National Curriculum as quickly as possible is great and this approach to implementing science has been common but leads to only short-term gains. Headteachers and staff need to treat the process of implementing the science orders as a whole, and lay down what is expected of each individual and the timescale involved in order to ensure not only coverage but quality teaching experiences for children.

Staff will need to set or be set clear and understandable criteria by which the success of each stage of the teaching and assessment processes can be measured. It is pointless to launch a new science programme unless everyone involved has a clear idea of what is expected from them as individuals, and as part of a teaching team.

Quality in any sphere takes time to achieve and to obtain quality you require a concise image of your goals. The teaching of science is no exception.

1

WHAT IS GOOD SCIENCE TEACHING?

Setting standards

All schools should take an overall approach to the planning of any curriculum area. At the outset of drawing up a science policy, all the teachers involved should have:

1. An agreed view of what they consider to be good science teaching.

This will help staff to see exactly what they and the school are aiming to achieve.

2. A clear understanding of what the statutory orders for science state and a thorough knowledge of exactly the science for which they have to ensure coverage.

This background will help staff when they draw up their Programme for Science and decide which aspect of science should be taught when, and whose responsibility it is to teach it.

WHAT IS GOOD SCIENCE?

Science in the primary school is concerned with children finding out about the world in which they live. It involves children developing knowledge and understanding of themselves and the world around them.

Science work in school helps children to develop their **knowledge and understanding** of scientific concepts enabling them to learn and use the **skills** associated with scientific methods of investigation. It fosters **attitudes** that promote scientific ways of thinking and working, such as curiosity, co-operation and sensitivity to living organisms. All these elements of learning are set out within the four Attainment Targets which make up the statutory orders for science.

KNOWLEDGE

Children need to be given opportunities to experience all areas of scientific knowledge which can be of use to them in their lives and in new situations. These areas of study are defined within Attainment Targets 2, 3 and 4.

SKILLS

'All children should have the opportunity to develop the skills of imaginative but disciplined enquiry' (National Curriculum Science Working Group, *Interim Report,* 1988). The skills of working as a scientist involve observation, hypothesis, design, investigation and the ability to draw conclusions and communicate effectively. All these aspects need to be developed in children, and are defined within Attainment Target 1 of the science curriculum.

ATTITUDES

Attitudes such as curiosity, perseverance and co-operation are developed through scientific activities.

2

WHAT ARE THE STATUTORY REQUIREMENTS?

Statutory requirements for schools when planning for science

To establish what the statutory orders for science in the National Curriculum are asking teachers to undertake with the children, staff need to consider the following aspects:

1. General introduction;
2. Attainment Targets;
3. Programmes of Study;
4. Statements of Attainment.

GENERAL INTRODUCTION

This is an introduction to the Programmes of Study set out at the start of each Key Stage. It is closely linked with Attainment Target 1 Scientific investigation and is particularly useful to teachers in helping to draw up the Science Policy as it outlines basic principles which they need to consider in their planning of science.

The following key extracts from the introduction set down clearly elements of what is considered good science teaching. Each point should be referred to when drawing up the Science Policy and followed through into the scheme of work.

AT KEY STAGE 1 Children should:

- Develop and use a variety of communication skills and techniques in obtaining, presenting and responding to information.
- Express their findings and ideas to other children, orally and through charts, models, actions and writing.
- Be introduced to books, charts, pictures, videos and the use of computers.
- Develop an awareness of the importance of science in everyday life.
- Use a variety of domestic and environmental contexts as starting points for learning science.

AT KEY STAGE 2 Children should:

- Have opportunities to develop and use communication skills in presenting ideas and in reporting their work to audiences, including other children, teachers, parents and other adults.
- Be encouraged to present information in an ordered manner.
- Be introduced to conventions used in diagrams, tables, charts, graphs, symbols and models.
- Participate in small group discussions and be introduced to books, charts and other reference sources.
- Use computers to store, retrieve and present their work.
- Develop an awareness of the important role of science in everyday life, including personal health and safety and the use of microelectronic devices to control appliances in the home.
- Be introduced to industrial as well as domestic and environmental contexts, as starting points for work in science.

ATTAINMENT TARGETS

The statutory orders for science are set down under four Attainment Targets.

AT1 This sets down the way in which children should work when undertaking investigations.

AT2

This is the study of the biology of plants and animals, classification, genetics and ecology.

AT3

This is the study of materials, rocks, minerals, soil and types of materials and their uses.

AT4

This is the study of physical processes including energy and forces, the Solar System and the Universe.

Strands

Each Attainment Target is divided into strands under which are set the Programmes of Study, which all schools should plan coverage of, and Statements of Attainment, to which all teachers will undertake assessments. The strands are set down to assist teachers in their planning and in their understanding of the progression through each Key Stage.

PROGRAMMES OF STUDY

Under each strand are set a number of related Programmes of Study which, when viewed together, illustrate clearly the progression through the Key Stages.

AT1 SCIENTIFIC INVESTIGATION

- strand (i) ask questions, predict and hypothesise;
- strand (ii) observe, measure and manipulate variables;
- strand (iii) interpret their results and evaluate scientific evidence.

This Attainment Target emphasises the need for children to be practically involved in investigations in order to develop their scientific knowledge and understanding. It requires children to observe and ask questions, predict and hypothesise, plan, design and carry out investigations, draw conclusions and communicate their findings. It centres around children working through a structural process which encourages them to work as a scientist.

Teachers should encourage children to plan and carry out investigations in which they work through the three strands.

The Programmes of Study for Attainment Target 1 are not grouped under the strand headings but show the areas that need to be covered by children at Key Stages 1 and 2.

The strand headings in Attainment Target 1 identify the areas that need to be assessed.

AT2 LIFE AND LIVING PROCESSES

- strand (i) life processes and the organisation of living things;
- strand (ii) variation and the mechanisms of inheritance and evolution;
- strand (iii) populations and human influences within ecosystems;
- strand (iv) energy flows and cycles of matter within ecosystems.

Attainment Target 2 is divided into four strands in which children should develop their knowledge and understanding of:

(i)   life processes and the organisation of living things

Children should find out about mammals and plants and the basic life processes. They should investigate factors that influence plant growth, and good health in humans.

5

(ii) variation and the mechanisms of inheritance and evolution

Children should consider similarities and differences between themselves, plants, animals and fossils. They should explore the wide variety of living things and be introduced to the fact that information is passed from one generation to the next.

(iii) populations and human influences within ecosystems

Children should study habitats and how animals and plants are adapted to live there considering human influences on the environment such as pollution.

(iv) energy flows and cycles of matter within ecosystems

Children should study the relationship between energy, the Sun, and plants and the significance of food chains. They should study the factors relating to decay and consider the effects of waste on the environment.

AT3 MATERIALS AND THEIR PROPERTIES

– strand (i) the properties, classification and structure of materials;
– strand (iii) chemical changes;
– strand (iv) the Earth and its atmosphere.

This Attainment Target is divided into four strands, only three of which are applicable to Key Stages 1 and 2.

(i) the properties, classification and structure of materials

Children should develop their understanding of the properties of materials comparing solids, liquids and gases and studying dissolving, evaporation and the separation of mixtures.

(iii) chemical changes

Children should explore the origins of materials and the effects of chemical changes when materials are heated or cooled.

(iv) the Earth and its atmosphere

Children should have the opportunity to study the weather and the seasons of the year. They should study local rocks, minerals and soils, how they are formed and the effects of weathering and erosion.

AT4 PHYSICAL PROCESSES

– strand (i) electricity and magnetism;
– strand (ii) energy resources and energy transfer;
– strand (iii) forces and their effects;
– strand (iv) light and sound;
– strand (v) the Earth's place in the Universe.

This Attainment Target is divided into five strands:

(i) electricity and magnetism

Children should develop an understanding of electricity and investigate the properties of magnetic and non magnetic material.

(ii) energy resources and energy transfer

Children should study energy, its sources, control and use. They should discuss the fact that not all energy resources are renewable.

(iii) forces and their effects

The children should study natural and manufactured forces, and their effects in everyday situations.

(iv) light and sound

Children should listen to the sounds in the environment and explore how sounds are made and heard, investigating light and shadow and how light and sound travel.

(v) the Earth's place in the Universe

Children should study the motion of the Earth, Sun and Moon and relate these to day, night, the seasons, year and day length, eclipses and phases of the Moon.

STATEMENTS OF ATTAINMENT

The Statements of Attainment give the teacher a clear indication of what a child should either be able to do or know within each strand.

Teachers need to consider the Statements of Attainment which relate to the strands they are planning to cover. There is a need for teachers to identify those statements they are planning to assess when they plan the work they are going to undertake in the classroom.

The Programmes of Study and the Statements of Attainment need to be used together to direct the teachers' planning of investigations and activities.

Schools have a responsibility to cover all the Programmes of Study. Planning to Statements of Attainment alone means that significant areas of the Programmes of Study could be missed. This would leave gaps in the children's scientific understanding and knowledge which may be vital for later work.

3

WRITING YOUR SCIENCE POLICY

What to consider when writing your policy

Once the staff have an understanding of what constitutes good science practice and are aware of what the statutory orders are asking them to cover, it is possible for them to make policy decisions relating to how and when the children will be introduced to the Programmes of Study for science.

Once policy decisions have been made it is easier for schools to identify how and when the various parts of the statutory orders will be taught.

WHAT MAKES A GOOD POLICY?

A simple guide as to what makes a good policy follows.

– Keep it short so that people will want to read it.
– Decide on the readership. For example, governors, teachers, parents, visitors to the school and inspectors.
– Ensure it is free from unnecessary scientific terminology, especially if it is going to be distributed to parents.
– A new teacher to the school, or supply teacher should be able to understand fully what is expected.
– Can you clearly see what you are setting out to achieve and will you be able to identify or measure your success?

Remember that a good policy is a management tool which will assist headteachers in bringing about change in the quality of education within their school.

A good indicator of how well a policy has been prepared and written is to ask whether a visitor to the school could read the policy and, if everybody were teaching science on that day, clearly observe aspects of the policy being addressed, worked towards or set down in the school's long-term plan.

DRAWING UP YOUR SCHOOL SCIENCE POLICY

Set out below are headings which form the framework for a school's Science Policy. Staff should be led through a series of staff meetings to address each of the headings and come up with an agreement to how information should be placed under each.

Under each of the headings below there is guidance on what issues might be discussed and suggested statements.

At the end of this section is an example Science Policy which schools might like to use as a guide. Schools may, of course, choose to adapt the example policy to meet their own situation and requirements.

THE PHILOSOPHY OF SCIENCE

What are the areas of knowledge, skills, attitudes and qualities which you feel need to be taught?

How will the general introduction to the Programmes of Study and the Programmes of Study for Attainment Target 1 help you to form a philosophy which promotes good science teaching and learning?

Does what you have written reflect what you agreed earlier was good science teaching?

SCIENCE IN THE NATIONAL CURRICULUM

Write a short paragraph which introduces the Attainment Targets.

How is the delivery of the strands to be paced and organised through the school?

When will each strand be covered?

Where will staff go to find out exactly what they are to cover and when?

Has the school drawn up a Programme for Science which sets down exactly when various aspects of the statutory orders are to be covered?

How will the children's skills of scientific investigation be developed through the school?

Will emphasis be given to any particular aspects of Attainment Target 1? How will children develop independence as set down in Attainment Target 1?

TEACHING STRATEGIES AND PLANNING

What guidance will teachers receive in planning a wide range of investigations and activities which are appropriate to the children's ability, knowledge and skills to be taught?

What approach will staff be required to take to the planning of science investigations and activities?

IN THE CLASSROOM

What organisational strategies will be encouraged in school to develop the children's knowledge, skills and attitudes?

In what ways will children be encouraged to report and record their work?

How will Information Technology be encouraged through the context of scientific investigations and activities?

EQUAL OPPORTUNITIES AND SPECIAL NEEDS

How will the school ensure that all children are given the same opportunities within their science work?

How will teachers plan to ensure that all children are having opportunities to work to their full potential whether it be at the lower levels or at the higher levels?

ASSESSMENT AND RECORD KEEPING

How will the school keep a record of when the Programmes of Study have been undertaken?

How will the assessment of science be carried out within the school?

How will the assessment of the knowledge aspects of the statutory orders be undertaken and how often?

How will the assessment of Attainment Target 1 be undertaken and how often?

What guidelines will be given to teachers on the keeping of evidence?

RESOURCES

How are the resources to be organised in school?

Are there to be resources kept centrally as well as some in the classrooms?

Is there to be a system adopted for the borrowing and return of central resources?

Will somebody be responsible for ensuring the smooth running of the system and replenishing of stocks?

How will the children be encouraged to use the resources?

EARLY YEARS

How will the very young children in school be introduced to science and the way of working?

SAFETY AND CARE

What safety aspects will you want to set down so that all present and future staff are aware of the policies on:

– use of electrical items;
– craft knives;
– experiments with food;
– other issues specific to your school.

How will the school encourage a safe approach in the teaching of science to ensure the well-being of all children involved?

How will the children be encouraged to handle and care for all living things?

Where is the Health and Safety Document kept?

REVIEW

How often will you as a school review your policy to update and refine it as appropriate?

HOW WILL THE POLICY SUPPORT SCIENCE TEACHING IN YOUR SCHOOL?

The writing and review of your science policy will encourage professional debate and increase the awareness of the staff as to what is required in their teaching.

It will help to develop continuity and progression and give guidance for maintaining and developing quality in teaching and in the children's work.

New members of staff should be able to look at the policy and know what is required of them in their teaching and planning.

It should also guide the headteacher, co-ordinator and any visitors as to what to look for when evaluating the quality of the science teaching taking place in the school.

The curriculum policy for science

The policy must cover the following.

– Reflect the ideals and philosophy which are promoted in the whole School Curriculum Policy for your school.
– Reflect the ethos which you hope to create for the teaching of science at your school.
– Guide the classteacher when planning a wide range of science activities for all children.
– Help to achieve consistency.
– Be short and written in a language everyone can understand, i.e. parents and governors.
– Reflect the work currently being undertaken in the classrooms or what teachers are working towards.
– Help the classteacher deliver the National Curriculum.

AN EXAMPLE SCIENCE POLICY

For the purpose of this publication we have written a school Science Policy based on the questions outlined on pages 10–12.

This example policy will form the basis around which we will map out the whole of the statutory orders for science (section 4). It will also influence how we will plan our investigations and activities to deliver the Programmes of Study for each strand and be represented throughout our Scheme of Work.

The ultimate test of the success of your policy is that it is reflected within the practice that is observed throughout your school.

THE PHILOSOPHY OF SCIENCE

For young children science is an introduction to the world of living things, materials and energy. It is a largely practical subject which develops a spirit of enquiry by encouraging curiosity and reason.

Scientists have revealed vast amounts of knowledge about our world by using skills of observation, prediction, investigation and interpretation.

Each child needs to enjoy the experiences associated with science by increasing and developing their knowledge and by starting to use the skills associated with scientific methods of investigation.

Working with others, learning how to persevere and learning how to ask questions are attitudes which encourage work to be carried out in a scientific way.

SCIENCE IN THE NATIONAL CURRICULUM

The three strands for Attainment Target 1, Scientific investigation, will be encouraged in every investigation or activity when appropriate.

Each child will be involved in at least one investigation every term which develops skills to:

- ask questions, predict and hypothesise;
- observe, measure and manipulate variables;
- interpret results and evaluate scientific evidence.

The twelve knowledge strands for Attainment Targets 2, 3 and 4 will be introduced once at Key Stage 1 and twice at Key Stage 2. At Key Stage 2 each strand must be introduced once every two years. The school's Programme for Science shows how these are mapped.

In addition, at Year 6, the children will have one session each week which involves revision and opportunities to recall and extend work from the previous year's programme.

TEACHING STRATEGIES AND PLANNING

It is important that the teacher identifies the most appropriate teaching strategy to suit the purpose of a particular learning situation.

There are a variety of ways in which the teaching may be effective and our school has a tradition for encouraging learning through investigation, with an emphasis on firsthand experience. It is, however, frequently acceptable to use demonstration, research, exploration, and teacher-led investigations when circumstances, resources and the needs of individuals and groups allow.

Teachers need to use their flair, enthusiasm, and professional judgement to identify the most sensible, enjoyable and safe methods for the work being conducted. The scheme of work provides suggestions to help in the selection of the most effective approach.

IN THE CLASSROOM

Teachers should look for opportunities to praise co-operation and safe, considerate behaviour.

Children are encouraged to work as individuals, in pairs, in groups and also as a whole class when appropriate.

The children are encouraged to use a variety of means for communicating and recording their work. The development of study skills and personal working styles is encouraged and respected. Participating as a speaker and a listener has a high profile at this school.

Information Technology, including computers, tapes and cameras, plays an important role in developing communication and data handling skills. Each child will use a computer at least once every half term for data handling and for interpreting results and findings. All the children will maintain a personal data base recording their growth and observations related to living things around the school. Each class will keep a weather record and all children will be involved in helping to compile and interpret the information gathered.

EQUAL OPPORTUNITIES AND SPECIAL NEEDS

Every effort is made to ensure that science activities and investigations are equally interesting for both boys and girls.

Children with special educational needs are involved in all work planned from the Programme for Science at an appropriate level which will help each child reach their full potential. Teachers' weekly plans show how the activities have been adapted or extended for children of different abilities.

ASSESSMENT AND RECORD KEEPING

The Programme for Science indicates both the Programmes of Study and Statements of Attainment to be considered for each term.

Teachers compile individual folders for each child. The folders contain both examples of work which have been assessed and notes relating to observations and conversations with the child.

Every child within the class should be assessed at least once a term for Attainment Target 1 either by using observation or by collecting written evidence.

In addition, updated assessments related to the strands indicated on the Programme must be made for each child at least once every term. Teachers could retain representative samples of class work.

RESOURCES

Every classroom has access to a resource area. The children are encouraged to choose from a range of equipment when designing investigations. Children are trained in the safe and considerate use of animals, plants and equipment and not to be careless with consumables and materials which are not easy to store.

The Science Co-ordinator is responsible for these areas. Any expensive items which are kept within the central store should be requested from this teacher.

EARLY YEARS

Reception children should be involved in any science activity which has an appropriate interest value and which has the capacity to excite and provide enjoyment.

The teachers of these very young children are encouraged to plan alongside their Year 1 colleagues and to use the Programme for Science as a structure for identifying activities which the children can undertake.

SAFETY AND CARE

The safe use of equipment is promoted at all times.

The ASE *Safety Policy* has been adopted by the staff and spare copies are available from the Co-ordinator for science.

The school's Health and Safety policy (available in the Staffroom, Office and Reception area) should be consulted for details regarding scissors, craft knives, electrical equipment, wet areas, heavy equipment and the use of tools.

Any animals, including insects, being used for study should be treated with respect and returned as soon as the activity is complete. For specific guidance related to work undertaken on 'decay' consult the ASE *Safety Policy*.

Leaves and berries of a poisonous nature should be avoided in classroom displays and their dangers made clear to the children.

REVIEW

This policy is reviewed by the staff and governors in the summer term. Parents are most welcome to request this document and comments are invited from anyone involved in the life of the school.

4

DRAWING UP A PROGRAMME FOR SCIENCE

Mapping the science curriculum

Once the policy is written schools will need to map out within a Programme for Science exactly how and when the various sections of the statutory orders for science are to be taught. This will help teachers in their planning, and enable headteachers to monitor coverage and aid the development of progression throughout the school.

The policy decisions you have made will allow you to establish how many times the children will experience the strands of the knowledge-based Attainment Targets and how your school will be approaching the teaching of Attainment Target 1.

Many schools will now have existing plans for science and these can easily be incorporated into a school Programme for Science.

A photocopiable blank Programme for Science sheet appears in Appendix 3, page 149.

AN EXAMPLE PROGRAMME FOR SCIENCE

The example Programme for Science has been set out to reflect the statements set down in the example Science Policy.

Key statements from the policy have been used to inform our decisions as to how and when the strands will be covered.

The example Programme for Science shows a two-year cycle in which all classes in the school will address the same strands at the same time. The first year of the cycle is Year A and the second is Year B.

It could be used by any school irrespective of the number of classes.

It could be used whatever the organisation of year groups within classes or whatever the elected style of delivery (subject or topic based).

As there are twelve strands which relate to the knowledge-based Attainment Targets, they are easy to divide up and fit into the three terms of a school year.

It is straightforward to follow when planning for assessment as the Statements of Attainment are directly linked to the Programmes of Study the teacher is delivering.

Within each strand (particularly at Key Stage 2) there are Programmes of Study which relate to particular Statements of Attainment. There are certain aspects of the Programmes of Study which are most relevant to those children working at higher levels. By planning to deliver strands in a two-year cycle the teachers will find it easier to plan for progression as they can identify those parts of the Programmes of Study for each strand which are most relevant to the abilities of the children in their classes.

USING THE EXAMPLE PROGRAMME FOR SCIENCE IN YOUR SCHOOL

KEY STAGE 1

Whatever your organisation of children for Key Stage 1 the Programme for Science could be used as it stands. The children will only have time to experience the strands for each Attainment Target once in the two years they are in Key Stage 1 so the whole Programme of Study has to be delivered in the term it is allocated to.

Programme for Science

A TWO-YEAR CYCLE (YEARS A AND B) FOR KEY STAGE 1 AND KEY STAGE 2 WHICH SHOWS:

– Each strand for Attainment Targets 2, 3 and 4 to be covered once at Key Stage 1 and twice at Key Stage 2;
– How each year a balance is delivered between Attainment Targets 2, 3 and 4;
– Statements of Attainment to be considered for each term.

AT1 Scientific investigation

strand (i) ask questions, predict and hypothesise;
strand (ii) observe, measure and manipulate variables;
strand (iii) interpret results and evaluate scientific evidence.

The three strands for AT1 will be encouraged in every investigation or activity when appropriate.
Each child will be involved in at least one investigation every term which involves skills from all three strands.

Ongoing work for all pupils in recording changes and observations for *weather, plants, animals and our growth.*
In addition, children at Year 6 will have one session each week for revision of areas related to the previous year's programme.

	Year A			Year B		
	Autumn	Spring	Summer	Autumn	Spring	Summer
	AT4 4.ii Energy resources and energy transfer	AT2 2.ii Variation and the mechanisms of inheritance and evolution	AT3 3.i The properties, classification and structure of materials	AT4 4.i Electricity and magnetism	AT2 2.i Life processes and the organisation of living things	AT3 3.iv The Earth and its atmosphere
	AT4 4.v The Earth's place in the Universe	AT2 2.iv Energy flows and cycles of matter within ecosystems	AT3 3.iii Chemical changes	AT4 4.iii Forces and their effects	AT2 2.iii Populations and human influences within ecosystems	AT4 4.iv Light and sound
	SoA (AT4 ii) 4.2b, 4.3b, 4.4b, 4.5b, 4.5c	SoA (AT2 ii) 2.1b, 2.2b, 2.4b, 2.5b	SoA (AT3 i) 3.1a, 3.2a, 3.3a, 3.4a, 3.5a, 3.5b	SoA (AT4 i) 4.1a, 4.2a, 4.3a, 4.4a, 4.5a	SoA (AT2 i) 2.1a, 2.2a, 2.3a, 2.4a, 2.5a	SoA (AT3 iv) 3.3c, 3.4d, 3.4e, 3.5d
	SoA (AT4 v) 4.1d, 4.2e, 4.3e, 4.4e, 4.5g	SoA (AT2 iv) 2.2d, 2.3c, 2.4d, 2.5d	SoA (AT3 iii) 3.2b, 3.3b, 3.4b, 3.4c, 3.5c	SoA (AT4 iii) 4.1b, 4.2c, 4.3c, 4.4c, 4.5d	SoA (AT2 iii) 2.2c, 2.3b, 2.4c, 2.5c	SoA (AT4 iv) 4.1c, 4.2d, 4.3d, 4.4d, 4.5e, 4.5f

Likewise, it will only be possible to assess against the statements for that strand during the term in which the strand is being taught.

At Key Stage 2 the teacher will need to make some decisions about which parts of the Programmes of Study will be taught each term, to whom they will be taught and what statements will need to be assessed.

These decisions will be influenced by the organisation of the class.

Mixed age-group classes

The programmes of Study at Key Stage 2 are extensive and as the children will have the opportunity to experience the Programmes of Study twice at Key Stage 2, the teachers in this phase need to decide which parts are most suitable to teach to particular year groups.

The decisions will be influenced by the Statements of Attainment, as these relate to certain parts of the Programmes of Study. For example, choose the Programmes of Study which relate to the higher level Statements of Attainment to undertake with the most able children.

As the Programme for Science represents a two-year cycle, the children will have an opportunity to experience the Programmes of Study again before moving to Key Stage 3 so areas which were not addressed and assessed the first time can be covered on the second coverage of the strand.

Year group classes

This situation is very straightforward. The teachers in Key Stage 2 simply need to decide which parts of the Programmes of Study, for the strand they are delivering, need to be taught by each teacher.

As in the previous example, this decision will be influenced by the Statements of Attainment. The Year 5 and 6 teachers will most probably be teaching those Programmes of Study which relate to the higher levels of understanding.

Assessment opportunities will relate to the parts of the Programmes of Study being undertaken by the teacher.

To help with this process of making decisions as to which parts should be taught the Scheme of Work lists Programmes of Study and their related Statement of Attainment on each page introducing the strand selected for study (section 5).

USING THE PROGRAMME FOR SCIENCE TO PRODUCE YOUR OWN SCHEME OF WORK

The example Programme for Science forms the basis of the Scheme of Work (section 5).

There are three possible ways in which schools may use this section of the book.

1. Schools may use the example Programme for Science and its associated Scheme of Work directly.
2. Schools may choose to adapt the Programme for Science to produce a Scheme of Work which is directly related to the particular situation of the school.
3. Schools may work through the stages identified and use the guidance provided to produce their own unique Programme for Science and associated Scheme of Work.

5

PRODUCING YOUR SCHEME OF WORK

Planning investigations and activities

Once a Programme for Science is available the teachers can use it as a foundation to plan a range of investigations and activities which will deliver and introduce the Programmes of Study for each strand to the children in their class.

WHAT SHOULD BE USED TO HELP IN PLANNING?

Teachers should always use the Programmes of Study as their starting point for planning. These are the learning experiences to which every child should be introduced. The Statements of Attainment should be used alongside the Programmes of Study to inform the teacher as to what the children should know or be able to do once they have been introduced to the Programmes of Study.

When planning investigations teachers should ensure that the investigation involves the children working through the whole process as identified by the three strands of Attainment Target 1. This will encourage the children to develop good practice in the carrying out of investigations, and to work in a logical manner, as well as giving the teacher an opportunity to make assessments against the three strands of Attainment Target 1.

The investigations reflect this approach (more guidance is provided in section 6, page 137).

WHICH TEACHING METHODS SHOULD BE USED?

There is not just one approach to the teaching of science. The teacher needs to make decisions about the most effective way of undertaking particular aspects of the Programme of Study while taking into account the needs of the children.

Different methods will include children being involved in investigations, exploration, demonstration, teacher-led investigation and research. It is expected that there will be a balance between the variety of teaching methods and on many occasions they will work together to provide the best learning experiences for children.

It is apparent, however, that no one technique in isolation can deliver the whole Programme of Study for all four Attainment Targets.

WHAT ORGANISATIONAL METHODS SHOULD BE USED?

The three main ways of organisation of the learning experiences for children in the classroom are: individual, group or whole class teaching. Teachers should be aiming for a balance of all three, as they all have a role within the teaching of science.

Individual teaching is ideal for the child who needs particular attention to help them develop understanding of a particular concept or to use a particular skill. There are times when whole class teaching may be needed when the teacher wishes to give information. Group work is very appropriate in the development of co-operative learning.

THE IMPORTANCE OF INVESTIGATIONS

Whilst it is very important for teachers to realise that there is not just one way of teaching science it is equally important for teachers to recognise the significance of children being involved in investigations.

The example policy and Programme for Science identify the fact that every child will undertake at least one investigation every term. This is a very important point to which attention should be drawn.

The skills associated with the development of Attainment Target 1 can be undertaken through a variety of approaches but it is only through the context of an investigation that the children have the opportunity to work through the whole process of asking questions, making predictions, planning and carrying out the investigation, recording their findings and drawing conclusions.

Likewise, children can only be assessed by the teacher against the Statements of Attainment for Attainment Target 1 when they are involved in investigations.

THE SCHEME OF WORK

For the purpose of this publication we have planned a Scheme of Work based on the Programme for Science found in section 4. Pages 25–136 contain investigations and activities which will help teachers deliver the Programmes of Study to the children at Key Stages 1 and 2.

The pathway through the pages is as follows.

1. Each year and term is identified and the strands to be covered in that term (based on the Programme for Science – see page 19).
2. The Programme of Study and Statements of Attainment for the first strand are identified for Key Stage 1.
3. A set of activities followed by an investigation which could be used to deliver the strand.
4. Possible assessment tasks which would allow the teacher an opportunity to assess against the relevant statements at this Key Stage are integrated into the activities and investigations.

This format is repeated for Key Stage 2.

After Key Stage 2 is the next strand from the programme.

The format described above is repeated for each strand through the Programme for Science.

This section therefore contains a complete Scheme of Work which includes activities and investigations suitable for children progressing through Key Stages 1 and 2.

The matrix on the following page summarises which activities can be used to assess each Statement of Attainment.

SUMMARY OF ACTIVITIES

ACTIVITIES

ATTAINMENT TARGET	KEY STAGE	STRAND	1	2	3	4	5	6	7	8	9	10	11	12
2	1	(i)	2.1a					2.3a						
		(ii)		2.2a		2.1b, 2.2b								
		(iii)				2.1b, 2.2b								
		(iv)		2.2d		2.2c								
	2	(i)		2.4a, 2.5c			2.3a							
		(ii)	2.5b			2.2.2b, 2.4b								
		(iii)		2.2c, 2.3b				2.4c						
		(iv)	2.5d	2.4d										
3	1	(i)		3.1a, 3.2a			3.3a							
		(iii)		3.3b			3.2b							
		(iv)	3.3c											
	2	(i)	3.3a			3.4a		3.5a		3.5b				
		(iii)		3.2b, 3.3b			3.5c		3.4c					
		(iv)	3.4d	3.5d										
4	1	(i)	4.2a	4.1a	4.3a									
		(ii)	4.2b		4.3b									
		(iii)	4.1b, 4.2c											
		(iv)	4.3c, 4.1c	4.3d	4.2d, 4.1c									
		(v)	4.2e	4.4e		4.1d								
	2	(i)	4.3a, 4.4a	4.2a	4.5a									
		(ii)	4.4b, 4.5b		4.3b, 4.5c									
		(iii)	4.3c, 4.4c			4.5d								
		(iv)	4.3d, 4.5f	4.3d, 4.5e	4.4d									
		(v)	4.5g	4.2e, 4.3e, 4.4e			4.4e							

Attainment Target 2 Life and living processes

Key Stage 1 strand (i)

Life processes and the organisation of living things

PROGRAMME OF STUDY

A Developing self-awareness

Pupils should find out about themselves and develop ideas of how they grow, feed, move and use their senses, and about the stages of human development.

B Plants – their growth and reproduction

They should be introduced to the main parts of flowering plants and investigate what plants need to grow and reproduce

C Health and exercise

They should be introduced to ideas about how they keep healthy through exercise, personal hygiene, diet, rest and personal safety; and to the role of drugs as medicines.

STATEMENTS OF ATTAINMENT

2.1a Be able to name the main external parts of the human body and a flowering plant (**A** and **B**).

2.2a Know that plants and animals need certain conditions to sustain life (**A** and **B**).

2.3a Know the basic life processes common to humans and other animals (**A**).

ACTIVITIES Life processes and the organisation of living things

PARTS OF THE BODY

RESOURCES   An outline or model of the human body

Class or group discussion   The children play a game of either placing labels on to the 'body' showing the main external parts, for example arms, legs, head (and associated features), feet, etc. Or the teacher places labels on the body incorrectly and asks the children to put them in the correct places. Children should not simply be able to say where different parts of the body are but should be encouraged to explain and investigate their functions in a simple way.

Classroom display   Make a classroom display by making an outline of the human body from different materials. Make labels for the different parts of the body. Add Velcro to the backs of the labels so that children can place them in the correct place on the display.

Extension Children could find out about how children and adults without arms and legs cope with their handicaps and could be encouraged to consider how these people can improve the quality of their everyday lives.

PARTS OF A PLANT

RESOURCES A collection of plants for children to examine closely

Teacher demonstration The teacher shows the children a flowering plant such as a daffodil or tulip and identifies the various parts of the flower, for example root, stem, leaves, flower. Children could then take a similar plant and draw in detail the parts identified. This could be done for a variety of plants. Children should be encouraged to look for similarities and differences between different plants.

ASSESSMENT Activities 1 and 2 combined provide opportunities to assess:
2.1a Be able to name the main external parts of the human body and a flowering plant.

KEEPING PLANTS ALIVE

RESOURCES Photographs, paper cuttings relating to famine

Class or group discussion Teacher discusses with the children why famine occurs and what reasons there are for famine in the world. Discussion could then centre around what we need in order to stay alive. Why do we need to eat and why do we need water? Is this the case for all animals?

Cross-curricular link Health education Children could then carry out research into what are healthy foods and what constitutes a healthy diet. Children could keep a diary of the food they eat in a week.

Practical Teacher could then pose questions about how and what plants need in order to grow which could lead to investigations set up by the children. (See investigation for this strand, page 29.) In order for children to achieve the Level 2 statement in this strand there is a need for them to be aware that not only do plants need water and food (just like animals) but also need light in order to sustain life.

SEEDS

RESOURCES A collection of seeds or fruits which can be opened to view seeds inside. (Good examples are: thistle, poppy, sunflower, dandelion, acorns and conkers)

Class or group discussion Discussion could centre around what plants the seeds come from and what is their purpose.

Practical The children could observe closely and make drawings of the seeds. Children find out how the seeds are transported and why. The most important aspect of this activity is to develop an understanding that most plants rely on their seeds to ensure the survival of their kind. The children could investigate seeds that are dispersed by wind, for example dandelion seeds. At the end of the activity the seeds could be planted to see how long they take to germinate.

ASSESSMENT Activities 3 and 4 combined provide opportunities to assess:
2.2a Know that plants and animals need certain conditions to sustain life.

5 WHAT IS EATING?

RESOURCES Biscuits

Practical Children watch each other eating a biscuit – what parts of our body can we see that are involved in the eating process? Where does the food go when we eat? Where does it go after it enters the mouth?

Find out where the food goes and what happens to it as it passes through the body.

6 THE BABY

RESOURCES A visit from a parent and baby, photographs of the baby showing clearly the changes that have taken place as it has grown older

Class or group discussion Discussion could take place about why humans have babies. Children talk about the baby and how it has changed since it was born (useful to have photographs of the baby at an earlier age). Discuss and write about the daily routines and needs of a baby. How do the needs of the baby compare with older children and adults, or even elderly people?

Discuss how the baby will change in appearance over the years. Children could compare how they have changed over the years – use old photographs. If birth weights and lengths are available, children could find out if the longest baby is now the tallest child etc.

RESOURCES It would be interesting to have a collection of pictures of animals and their young: mammals, insects, reptiles etc., and encourage the children to find out about how the young develop and are looked after by their parents

7 MOVEMENT

Cross-curricular links Physical education Children during a physical education lesson are encouraged to show movement in a variety of ways, running, jumping, skipping, rolling etc.

Describe which parts of the body are used when they move – and are all these parts used whatever the type of movement, for example running, jumping etc.? Children could then discuss how their bodies feel when they move or do exercise.

Children could collect pictures of various athletes and sports people involved in movement. They could be introduced to the idea of muscles, bones and joints helping us move.

8 SENSES

RESOURCES Set up an 'interest' table with a collection of materials and equipment that encourage the children to use their senses. Collect objects that children can describe by smelling and feeling. Be careful about placing objects on the table that children will have to taste

Sound Children play a game in which they are blindfolded – sounds made – can the child guess what the sound is?

A sound from a different direction – can the child indicate which direction the sound is coming from?

Tape recordings of sounds around the home – can they be identified?

Sight Set up an eyesight test so that children can test their own sight and learn how distance affects how well they can see and recognise features. Does it make a difference which eye they look through? Is it better to use both eyes? What difference does it make when they only look through one eye?

Taste and smell Children play guessing games to see if they can recognise different materials by using just taste and/or smell.

Touch Mystery object sack. Children play the game – by putting their hands in the sack and telling each other what they can feel. Look closely to see what parts of the hands are used for touching. Children could create a 'feely board' of pleasant objects to touch – children describe why they like to touch the objects.

 KEEPING OURSELVES CLEAN

RESOURCES A collection of items used to keep ourselves clean

Class discussion Why is it important to keep clean?

Children discuss what things we use to keep us clean and tidy and how and where these things are used. Pose questions about the need to keep ourselves clean.

Practical Are there parts of the body that get dirtier than others? Children could investigate which soap is best for washing hands. Does the temperature of water make any difference when washing – try cold water, hot water.

 KEEPING OURSELVES FIT

RESOURCES A collection of sporting pictures and sports equipment

Class discussion What sports are the pieces of equipment designed for? The children could discuss the reasons why people play sport. A survey could be carried out throughout school to find out which is the most played sport in school.

How do you feel after exercise? How does your body react to exercise? Discussion could be based around heart beat, breathing rate, temperature. The children should have explained to them in simple terms what is happening in the body during exercise.

Practical The children could investigate heart rate/breathing rate after certain types of exercise.

Children keep a diary of time spent exercising and time spent resting.

ASSESSMENT Activities 5 to 10 offer opportunities to assess:
2.3a Know the basic life processes common to humans and other animals.

The emphasis of this statement is on the children knowing that animals and humans need to feed, breathe, move and reproduce.

 SAFETY AT HOME

Class discussion A discussion about safety in the home and outside.

Practical Children carry out a survey into how and where most accidents occur. Which people are most at risk? Children could put together a safety campaign in which reference is made to which household equipment should be kept away from children and how children and adults can be safer around the home and outside. What ways have been devised to prevent accidents around the home?

 MEDICINES

RESOURCES | *George's Marvellous Medicine* by Roald Dahl

Class discussion | Discuss with the children why we have medicines. What type of illnesses have medicines been taken for in your class? Discuss the safe use of medicines – why is it important?

INVESTIGATION

Life processes and the organisation of living things

"**What things are needed for a seedling to grow?**"

Starting point
A packet of seeds or a tray of seedlings.

Observing and asking questions
Discussion between teacher and children about what seedlings are and where they come from. Also discuss what do you need to make seedlings grow. Reference could be made to the growth medium, water, light, plant, foods and the temperature and why they are important. Observe a variety of plants in different states of health.

Children may ask
What type of soil is best for growing seedlings?
Do seedlings need water in order to grow?
What type of water is best for a seedling to grow?
Do seedlings need light in order to grow?

Encourage the children to identify one thing to investigate at a time: for example, growth medium.

Predicting and hypothesising
Before carrying out the investigation always encourage the children to make a prediction and give a reason for their prediction.

The compost will be best for growing seedlings because it contains food that the seedling can use as it grows.

The seedling which has no light will not grow as well as the seedling with light because seedlings need light to help them make food.

Designing and planning the investigation
When the children have decided what to investigate encourage them to decide which things they are going to change and which things they are going to keep the same.

Growth medium Children have a minimum of three different growth media, for example; soil, compost and sand, and sand and investigate which is best for growing seeds. Everything else within the investigation should be kept the same for example, quantity of growth medium, amount of water given, type of seed, place kept.

Water Children investigate one seed with water and one seed without – everything else should be kept constant.

Resources
- tray of seedlings
- either broad bean, cress or pea seeds
- different growth media: e.g. variety of soils, compost, sand
- water – different types: e.g. rain, tap, salt, distilled (obtainable from a garage)
- plant pots/ seedling trays
- rulers, measuring cylinders, scales

Light Children investigate one seedling with light and one seed without – everything else should be kept constant.

Children could investigate different amounts of light, for example, dark, shaded, full light. All other things should be kept constant.

Children then decide what equipment they will need in order to carry out their investigation. Teacher offers support and guidance in the selection of equipment and shows the children how to use measuring instruments if needed.

Children need to decide how they will judge whether the seedlings are growing successfully or not. The seedlings could be measured (using standard or non-standard measures); they could be sketched and coloured to show healthy colours.

The children set up their investigations ensuring that everything is kept the same apart from the aspect they are investigating.

Recording

Decisions need to be made about how often they are going to record their observations and in what form – they are encouraged to carry out their observations accurately and record them carefully.

The children can record their findings in a table, chart or graph. Information Technology may be used to help them. Data handling packages such as *Junior Pinpoint* or *Data Sweet* (Archimedes), or an art package such as *Flare* could be used for recording results.

Drawing conclusions

Children should look carefully at their results and give an explanation as to what has happened and why.

The seedling grown in the compost grew much better than the seedling grown in the sand because there is much more goodness in the compost which the seedling used to help it grow.

The seedling grown in the dark lost its green colour and became long and thin. This happened because the seedling needs light to help it stay green and healthy and I think it was trying to find some light, that's why it grew long and thin.

Encourage the children to try and explain what has happened, rather than simply to report what happened. Wherever possible encourage the children to relate their findings to their original prediction.

Attainment Target 2 Life and living processes

Key Stage 2 strand (i)
Life processes and the organisation of living things

PROGRAMME OF STUDY

A Organ systems and life processes

Pupils should be introduced to the major organs and organ systems of mammals and flowering plants. They should explore some aspects of feeding, support, movement and behaviour in relation to themselves and other animals. They should explore ideas about the processes of breathing, circulation, growth and reproduction.

B Plant growth

They should investigate the factors that affect plant growth, for example light intensity, temperature and water availability.

C Health, drugs and medicines

They should study how microbes and lifestyle can affect health, and learn about factors that contribute to good health including defence systems of the body, diet, personal hygiene, safe handling of food, dental care and exercise. They should be introduced to the fact that while all medicines are drugs, not all drugs are medicines. They should begin to be aware of the harmful effect on health resulting from an abuse of tobacco, alcohol and other drugs.

STATEMENTS OF ATTAINMENT

2.2a Know that plants and animals need certain conditions to sustain life (**A** and **B**).

2.3a Know the basic life processes common to humans and other animals (**A**).

2.4a Be able to name and locate the major organs of the human body and of the flowering plant (**A** and **B**).

2.5a Be able to name and outline the functions of the major organs and organ systems in mammals and flowering plants (**A** and **B**).

ACTIVITIES Life processes and the organisation of living things

THE EFFECTS OF EXERCISE ON THE BODY

Class or group discussion

Discussion about how the body reacts to exercise after a physical education lesson. Children could discuss why they feel their heart beats faster when they do exercise. They could look for bodily changes during exercise, for example, breathing rate, skin colour, sweating and heart beat rate. Research work could be done to find out what is happening in this situation.

Practical

Children could investigate the differences between each other's heart beat rate and breathing rate after exercise. They could investigate different types of exercise and find out which is the most strenuous.

 THE MAJOR ORGANS

RESOURCES

A model of the human body

Research task

Find out about the major organs of the body. Where are they situated and what are their functions? How can we keep them healthy? Knowledge of the major organs needs to be developed with children at KS2 into a fuller understanding of how the organs work within systems in the body. For example, the circulatory, digestive and nervous systems. Local hospitals and healthcare centres could be contacted for information.

Opportunities should be provided for the children to compare organ systems in humans and mammals, and to identify any similarities and differences. The local vet may be able to help with information or visit the school to answer children's questions.

 THE HUMAN HEART

RESOURCES

A model of the heart

Class discussion

Discuss what the heart does. Research how the heart works within the whole circulatory system. Refer to arteries and veins.

Children could find out about the importance of exercise in keeping the heart working effectively. Contact the British Heart Foundation for information and details.

Practical

Children could investigate the effects of exercise on heart beat rate.

 PARTS OF THE PLANT

RESOURCES

A collection of plants, in flower if possible

Class discussion

Children find out the names of each plant through discussion with their teacher.

Practical

Examine closely the flowering parts of each plant using magnifying glasses and hand lenses. Children should draw what they see. Using reference material the children could find out the names of the different parts of a plant. They could also find out the function of each part. Children should produce a drawing of a flower. The drawing should be labelled with the following: petals, stamen, carpel, anther, ova, stigma and sepal. The children should be familiar with the function of each part.

ASSESSMENT

Activities 1 to 4 combined with activities 1 and 2 at KS1 will provide opportunities to assess:
2.4a Be able to name and locate the major organs of the human body and of the flowering plant.
2.5a Be able to name and outline the functions of the major organs and organ systems in mammals and flowering plants.

In order for children to achieve the above Statements of Attainment they should complete activities 1 to 4.

5 OUR TEETH

RESOURCES

A visit from the school dentist. Collect information relating to the job of the dentist

Class discussion

Children could find out about the nature of teeth. Why they are different shapes and what their different functions are.

How should teeth be looked after and why is it so important to do so? What things can cause decay and damage to gums and teeth?

Practical

Put together a dental care advertising campaign that informs people of the importance of careful dental hygiene.

6 DISEASE

Class discussion

Discuss diseases, how they are spread and how we become infected.

Practical

Children should research harmful germs: viruses and bacteria.

What common diseases are caused by these agents? Children could also research into diseases that can be caught only in foreign countries.

Children find out how the body combats disease. Contact local doctors, healthcare centres and hospitals for information on how our bodies are helped to fight disease. Mention inoculation and the use of vaccines.

7 AN ANTI-SMOKING CAMPAIGN

Class discussion

A discussion should be held between teachers and children on the issues against smoking.

Practical

The children could research tobacco smoking and its harmful effects and why people smoke. Questions to be answered could include:

– Where does it come from?
– What effects does smoking have on the body?
– Who does smoking harm most of all?
– Children could contact the local health authority for information and carry out surveys to obtain people's opinions.

Fact sheets and brochures could be put together to dissuade people from smoking, and placed in a doctor's surgery.

Information gained could be used to inform a debate in class over whether or not smoking should be allowed in public places.

8 HARMFUL DRUGS

This activity could follow on from the anti-smoking campaign.

Class discussion

The teacher should develop the children's understanding of what drugs are. Children should find out what the differences are between drugs which are medicines and drugs which are not, through discussion.

Which drugs are illegal to use, buy and sell?

Useful publications to help teachers handle this very sensitive area of the Programme of Study include:

The Good Health Project, T. Williams, N. Welton and A. Moon, Forbes Publications
My Body Health Authority Project, Heinemann Educational
Health for Life 2, Health Education Authority's Primary School Project, Nelson.

ASSESSMENT Activities 5 to 8 whilst contributing towards a child's overall understanding of life processes offer no opportunities for assessment against Statements of Attainment.

X-RAYS

RESOURCES A collection of X-ray photographs

Practical What can the children see on the X-ray photographs? The children could be encouraged to use a picture or model of a skeleton to see if they can name the various bones. Explanation could be given as to what purposes the bones of the skeleton serve. Children could find out about the joints of the body and identify where hinge, ball and socket, and universal joints are found.

The emphasis of this activity is to develop a child's understanding of the function of the skeleton and joints in producing movement in the human body.

The children could then investigate what helps the joints to move – the use of muscles – identifying pairs of muscles that work together.

HEALTHY FOODS

RESOURCES A selection of natural and processed foods

Practical Children should group the foods into animal products and those that are grown.

Using food labels the children could investigate which of the foods have substances added to them.

Using reference material children could help to identify constituent parts of foods, for example, protein, fat and carbohydrate – and answer questions such as:

– What jobs do they do in the body?
– Which foods do people think are good for them? (Give reasons.)
– Which foods should be eaten less frequently? Why?
– What is meant by a balanced diet?

ASSESSMENT Activities 9 and 10 will help to contribute towards a child's understanding of:
2.3a Know the basic life processes common to humans and other animals.

INVESTIGATION Life processes and the organisation of living things

"What do plants need to stay alive and healthy?"

Resources
- variety of seedlings and plants to observe
- bean and pea seeds
- growth media: soil, compost
- plant pots
- seed trays
- rulers
- measuring cylinders
- scales

Starting point
Look at a selection of different types of plants, some healthy and some unhealthy.

Observing and asking questions
Describe the plants and the differences between them. Discuss with the children:

- What do plants need to keep healthy?
- Why do you think some of the plants are looking unhealthy?
- Why do plants need light, water and food?

The needs of plants could be researched.

Children may ask
Does the volume of water given to a plant affect its growth?
Does the direction of light affect how well a plant grows?
Does the quantity of soil a plant is grown in affect its growth and its health?

More able children may wish to investigate two things, for example, does (i) the type of water or (ii) the quantity of water have most effect on how healthy a plant remains?

If this investigation is carried out the child needs to perform two tests, one for each type of water and the other for quantity of water. The child should then draw conclusions based on the results of both tests.

Predicting and hypothesising
Before carrying out an investigation always encourage the children to make a reasoned prediction. Children working at level 5 in AT1 should be volunteering scientific reasons for their predictions. This will mean that they require background knowledge on what plants need to stay alive and healthy.

The children may predict the following at level 3 (strand (i)) AT1:

The seedling that we are giving most water to each week will not grow so healthily because I've noticed that when our plants are over-watered at home they sometimes die.

At level 5 (strand (i)) AT1 the prediction may be:

The seedling that we are giving the least amount of water to each week will grow most healthily because seedlings need some water but not too much to help in the process of photosynthesis.

In this example it would be necessary to talk to the child about what they understood by photosynthesis and what was taking place in the plant for this to occur.

Designing and planning the investigation
When the children have decided what to investigate, encourage them to decide which variables they will change and which they will keep constant.

Water volume Children investigate a minimum of three different volumes of water. The type of plant, the composition and quantity of soil, light and temperature should be kept constant. The children should be encouraged to test different volumes of water on more than one plant (five plants for each volume of water). Ensure that the volumes selected by the child are varied enough to give different results.

Light direction A plant should be exposed to light from above, from below or from all sides. Children should investigate a minimum of three directions. All other factors in the investigation should be kept constant.

Mass of soil Ensure that the children use the same soil composition within their investigation. They should use a minimum of three different soil masses. The same type of plant should be grown.

Small pea or bean seedlings should be used. They can be grown quickly and in large quantities for the purposes of the investigation.

Children should decide what equipment they need before beginning their investigation. They should also consider where the investigations should be sited in the classroom to avoid interference or damage.

Children should select and use appropriate measuring devices. The children need to decide what observations and measurements should be made, and how often. A suitable method of presenting their results should also be planned. At KS2 the children should be working more independently in the design and implementation of their investigations. Their tests should be fair within the restrictions of a primary classroom.

Recording
The children should record their observations clearly. Efforts should be made in KS2 to encourage the children to quantify their observations using standard measurements.

Presentation of results in tabular or graphical form should be encouraged. Children should use IT whenever possible as a means of recording and presenting their findings.

A diary may be useful in which the children could draw the plants and record observations as their investigations progress. Consideration needs to be given as to how the findings will be presented to the other children in class.

Concluding
Children examine their results and report what happened and explain why. The children should be able to say whether or not the investigation they carried out was fair. If it was not, can they explain why?

If possible the children's findings should refer to the data they have collected to add support to their conclusion.

Children should ask: *Was the prediction I made accurate?*

The children should evaluate their investigation and pass comment on whether it was successful and if there are things they would change if they were to do it a second time.

A common fault in preparing for this investigation is to grow too few seedlings for each variable being tested. The children should be encouraged to recognise that when investigating how well plants grow, germinate or stay healthy, it is better to use more than one seedling for each variable being changed in order to obtain more reliable results. For example, the children are investigating how the quantity of water a seedling receives each week affects its health. The children should have at least five seedlings for each volume of water and then a range and average growth can be calculated.

This point is unlikely to be recognised independently by the children and should be introduced as a teaching point.

Attainment Target 2 Life and living processes

Key Stage 1 strand (ii)
Variation and mechanisms of inheritance and evolution

PROGRAMME OF STUDY

A Similarities and differences

Pupils should consider similarities and differences between themselves and other pupils and understand that individuals are unique.

B Animal and plant life and extinction

They should have opportunities, when possible through firsthand observation, to find out about a variety of animal and plant life and become aware that some life forms became extinct a long time ago and others more recently.

C Living things and their features

They should sort living things into broad groups according to similarities and differences using observable features.

D Caring for living things

Over a period of time, pupils should take responsibility for the care of living things, maintaining their welfare by knowing about their needs and understanding the care required.

STATEMENTS OF ATTAINMENT

2.1b Know that there is a wide variety of living things, which includes humans (**B** and **C**).

2.2b Be able to sort familiar living things into broad groups according to easily observable features (**B** and **C**).

ACTIVITIES Variation and the mechanisms of inheritance and evolution

LOOKING AT MINIBEASTS

RESOURCES This activity is easier to undertake if the school has an environmental area. Children will need to have access to the correct equipment for collecting and observing minibeasts. Collect observation containers, magnifiers, hand lenses and an old fish tank or shoe boxes for creating suitable environments

Class discussion The children could be asked to share their knowledge of where minibeasts are to be found around the school and in their neighbourhood. A class discussion could also invite the children to consider places where they think similar creatures might be found and to talk about the conditions which they seem to favour.

Where can we find insects around the school? Under stones, bark, walls and hedges? Observe them carefully in their environment.

Do they all like living in the same places?
How do they move?
What special features can you see? How do they protect themselves?
How do you think they see, hear, smell and feel?

How could you collect these creatures? How must they be handled? How are they to be treated if we are going to look at them closely?

When observed closely we need to be looking for differences and similarities between the animals in the collection.

Do you know the proper names for some of these animals?
Use reference books to help us find out the names of others.

The children could also make up names for individual or groups of insects. This would help the children appreciate the classification of animals into species and encourage close observation. Ask the children to classify the creatures according to observable features, for example, number of legs, colour and ways of moving.

Practical Drawing the animals and recording the place where each animal was discovered may help the children identify further places inhabited by a particular type of insect.

Do all little creatures like the same conditions?

Which of the creatures appear to prefer damp, dark places? (Woodlice are ideal for young children to observe and study.)

The children should be encouraged to devise simple investigations to test their ideas. A simple environment with contrasting areas of damp and light will help to confirm some of the predictions raised in the class or group discussion.

The children might then create a home for woodlice, in an old shoe box or something similar, in the classroom based on the knowledge which they have gained.

The importance of care and safe return of all animals after a short period of study should be stressed.

2 TREES

RESOURCES Trees and shrubs

Practical Groups or individuals could adopt a tree or shrubs within the school grounds, at home, or in the immediate locality of the school and start a diary containing brief observations and comments made over a long period of time. (The data could be kept on a child's personal data disk and be updated at regular intervals during their time at the school.)

The children could examine the leaves of the tree and consider how they are similar or different to those on plants.

The bark, branches and shape of the tree could be examined and the children could record the colours and texture of the tree. Leaves could be picked to press in work books. Measurements, photographs, sketches and written observations should all be added to the diary to record the seasonal changes and growth of the tree.

How often will you return to look for changes on the tree?
How will you show your findings?
How could you inform the other children about your observations?
Are there other plants which appear to change in similar ways? Which other shrubs and trees appear to change little between the seasons?

Ask the children to group the plants they have observed according to observable features such as whether they shed their leaves in winter, the shape of the leaves and the way the leaves are arranged on the branches.

3 WHERE PLANTS GROW

Practical Looking at plants around the school.

Where can plants be found around our school? Are there any plants growing in unusual places (in walls, between paving stones)?

How might they have been planted?
Why do they seem to grow quite well in these places?
Which places around school seem difficult for plants to grow? Can you suggest reasons why?

4 THE BIRD TABLE

RESOURCES A bird table, or feeding area

Deciding on the most suitable location for a bird table and its design are likely to create considerable interest in the project amongst the class.

Practical The children could investigate which foods different birds like.

Learning the names of common birds could be encouraged with reference material and by the children sharing their own knowledge. Binoculars and sketching materials left near the observation area could encourage a routine for noting behaviour and characteristics.

How are birds different related to their size, colour, beaks, feet and feeding habits?

These classifications could result in the children making their own charts based on first hand observations.

The number of birds visiting the feeding area could be recorded and graphs or frequency charts drawn to represent the data. Areas around the school where birds appear to congregate could be identified.

Children could be encouraged to reason why certain localities have a large bird population and some a smaller population.

The RSPB is pleased to send a free teacher's pack with many ideas for bird study projects and other resources. Copies can be obtained from:

The Education Department
Royal Society for the Protection of Birds
The Lodge
Sandy
Bedfordshire SG19 2DL
Tel: 0767 80551

DINOSAURS

RESOURCES A collection of model dinosaurs

The collection gives the opportunity to ask the children to identify the differences and similarities with creatures living today.

Many children will know and be keen to offer the names of the dinosaurs.

Class discussion The discussion should introduce the term 'extinct'.

The fact that these creatures lived so long ago may need to be emphasised.

The size, shapes and characteristic features of the different plant eaters and carnivores could be introduced with the models. How these different features might have helped or hindered their survival in the prehistoric world could be discussed.

The immense size and shape of certain dinosaurs could be drawn in chalk on the playground. (Older children would enjoy leading this task.)

How do we know what these creatures looked like? A reference book with pictures of fossils or reconstructed skeletons could prompt discussion.

Practical Individuals or groups could create a data base of prehistoric creatures. Each data base could be accompanied by a list of questions for friends to answer, for example: *Which was the biggest land creature?*

Which creatures from long ago still live on Earth today? (Crocodiles and sharks.)

What possible reasons can the children offer to explain the extinction of the dinosaurs?

Why have some prehistoric species survived? A child's hypothesis backed up with reasons based on the evidence contained in reference material should enjoy a high profile in the class discussion.

 CARE OF ANIMALS

RESOURCES *Pelican*, B. Wildsmith, Oxford University Press

Practical This popular book takes about twenty minutes to read aloud and is an excellent starting point for many of the questions listed below:

– How was the pelican born?
– Which other animals lay eggs?
– What does an egg need in order to be safe and to produce a baby animal?
– How do different animals protect their eggs?
– Which skills do baby animals learn quickly from their parents?
– What problems occur when young, wild animals are taken from their parents? When can they be returned safely to the wild?

An incubator can be loaned to the school and the interest and opportunities for firsthand observation are very exciting for young children. The care of the chicks does, however, need to be considered before launching into such a venture.

There are a number of organisations directly involved in helping to protect endangered species. For example, The World Wide Fund for Nature may be able to provide information on those creatures most at risk.

WATCH, the junior wing of the Royal Society for Nature Conservation, can be contacted at:
22, The Green
Nettleham, Lincoln LN2 2NR

7 LOOKING AT OURSELVES

RESOURCES

Photographs taken by the school photographer

Class discussion

From looking at their class photographs the children could discuss:

In which ways are we the same? (Parts of the body.)
In which ways are we different? (Size, weight, hair and eye colour.)

The photographer's contact prints could help with a display to illustrate observations.

ASSESSMENT

Activities 1 to 7 together offer opportunities to assess:
2.1b Know that there is a wide variety of living things, which includes humans.
2.2b Be able to sort familiar living things into broad groups according to easily observable features.

INVESTIGATION   Variation and the mechanisms of inheritance and evolution

"In what ways are the children in our class different?"

Resources
* bathroom scales
* tape measures
* rulers
* stop clocks

Starting point
Collecting data about ourselves.

Encourage the children to discuss and consider ways in which we are different. (Some children will be sensitive about their size, shape and weight and the working groups may need to be chosen with care.)

Observing and asking questions
The children could collect data about each other and start their own personal information file. (A personal data file could be compiled on disk by the child and updated at regular intervals during their time at the school.) Information could be collected on the children's height, hair and eye colour, hand spans, stride length, leg measurements, etc.

Using the data file children could answer questions such as who is the tallest in the class? As an alternative everybody's height could be recorded by using life size 'cutouts'. The children could be involved in arranging the information to show the order of height within their class or group of friends.

The teacher will know which children need to use, or refine, their use of standard measures. Younger children may wish to suggest some non-standard measures which would be appropriate.

Similar activities could be encouraged to collect data about the size of feet, thickness of arms or length of legs. All of the above activities should be carried out to develop children's awareness of the fact that we have physical differences.

Children may ask
Who has the strongest arms?
Who has the strongest legs?
Who can long-jump the furthest?

Who can jump the highest?
Who can run the quickest?

Predicting and hypothesising

Because the children have carried out considerable research and have collected data on the physical differences between them they will be able to make informed predictions about the questions they have asked.

Their predictions will be based around the information they have collected and a personal view about what makes a 'a good jumper', or a 'strong person'. Predictions could be things like: *I think that the tallest person in the class will be able to long-jump the furthest because their long legs will stretch further than anyone else's* or *I think the person with the thickest arms will have the strongest arms because they have more muscle than the others.*

Designing and planning the investigation

Encourage the children to decide what variables they will change and which they will keep constant. In all these investigations the changing variable is the individual child.

Strength of an arm or leg Each child will need to be investigated. All other aspects of the investigation need to be constant. For example, the weight to be held, the arm or leg to be used, the height the weight is held at and the position of the person being investigated (sitting or standing). Great care should be taken so that children do not strain or damage themselves.

The format identified above is applicable to all similar investigations.

Height and length of jump The starting point and position of the jump and where the distance and height are measured should be kept constant. A decision should be made as to how many attempts at the jump each child will make.

Running velocity Length of the course, starting position and clothes worn by the children should be kept constant. A decision should be made as to how many runs each child will make.

Recording

Pictograms, block graphs, line graphs or sets could be used to record the data. Photographs to compare the tallest and shortest members of the group would make a good record.

Drawing conclusions

The most interesting point of the investigations will come when children look back at their original predictions to see if they were accurate. It will also be very interesting to hear what explanations they give for the data they obtained. These explanations and conclusions will need to be handled carefully and sympathetically to avoid the element of competition and children feeling sensitive about their performances.

Another challenging development of the investigations is for children who are most able to look for answers to questions that require two sets of data they have collected to be analysed.

Further questions that may be posed by the children could include:

Does the tallest child have the biggest hands?
Did the children with the longest legs perform better in the long-jump investigation than those with shorter legs?
Do children of similar height wear the same sizes of shoe?

Attainment Target 2 Life and living processes

Key Stage 2 strand (ii)
Variation and the mechanisms of inheritance and evolution

PROGRAMME OF STUDY

A Plants and animals; the similarities and differences

Pupils should investigate and measure the similarities and differences between themselves, animals, plants and fossils.

B How plants and animals become fossils

They should be introduced to how plants and animals can be preserved as fossils.

C Using structural features to identify species

They should have the opportunity to develop skills to identify species of animals that occur locally by observing structural features and making and using simple keys.

D Genes and inheritance

They should be introduced to the idea that information is passed from one generation to the next.

STATEMENTS OF ATTAINMENT

2.2b Be able to sort familiar living things into broad groups according to easily observable features (**A** and **C**).

2.4b Be able to assign plants and animals to their major groups using keys and observable features (**A** and **C**).

2.5b Know that information in the form of genes is passed on from one generation to the next (**D**).

ACTIVITIES

Variation and the mechanisms of inheritance and evolution

INHERITANCE

RESOURCES An old class photograph

Class discussion The discussion is certain to promote the question: In what ways have we changed? (For example, height, weight, hair and other features.)

Comparing a personal photograph with that of a parent may encourage the child to consider how similar or different they are to other members of their family. (For example, hair, eyes, stature and features.)

Identifying similarities could be incorporated into a family tree, and the question asked: What are the likenesses in your family?
Which family features have been passed from one generation to the next?
Explain that these features are passed on in the form of genes.

ASSESSMENT **2.5b** Know that information in the form of genes is passed on from one generation to the next.

Fossils

RESOURCES A collection of fossils (could be models). Reference material

Practical Look closely at the fossils. Use hand lenses to get a more detailed view. Examine and sketch the ammonite, trying to put in as much of the detail as possible.

Cross-curricular links What do you think the fossilised animal looked like when it was alive? From
History the fossil, what picture can you build up of the shape and size of your animal? How do you think it might have lived long ago?

The children could use reference material to find out how long ago these creatures lived and where their fossils are commonly found today.

Teachers could ask: How were fossils formed? Which prehistoric creatures would have seen ammonites and possibly fed on them? Which living things today have similar features to the ammonite? Why do you think the ammonites became extinct?

Practical Use a variety of materials to make representations of fossils. Use shells, plants and other suitable objects pressed into Plasticine or clay or plaster of Paris.

Prehistoric shellfish, such as Devil's Toe Nails (*Gryphaea*) and Belemnites, can be found in some local gravels. Can the children offer reasons why these fossils are found commonly in locations now far from the sea?

Where plants grow

RESOURCES Different plants growing in a wild area or at a site where a variety of plants growing in different locations can be observed easily

Practical Which of the plants can the children name?

Which of the plants appear to have a variety of names offered by the children? Encourage the children to make up a name for each different plant based on observable features which can be used until reference books reveal the correct names.

Which of the plants only seem to grow well in wild areas? Which of these would be unsuitable for a garden or play area? (For example, nettles.)

The children could look carefully for clues to indicate how the different plants may be suited to their environment. For example protection, light and moisture. Encourage the children to link their observations with their knowledge of what plants require in order to thrive.

Cross-curricular links This activity can be undertaken in the local environment or on a field trip.
Geography Each child or group could choose one type of plant to study and indicate on a map where it grows best. Children should draw upon their knowledge of

maps. Would somebody who did not know the area be able to find the plant you have been studying from the details given on your map?

Art Children should be given experience of sketching and making notes at a site. Take to the study site materials for children to make detailed sketches of the plants they are studying.

PRODUCTS FROM ANIMALS

RESOURCES A garment made from animal skin or an imitation skin of man-made fibres

Class or group discussion Discussion could be encouraged around the following questions:
Which animals are hunted and why?
Why are they hunted?
What effect does this have on their ability to survive as a species?
Which animals have become rare or extinct because of hunting?
Which animals have become rare or extinct for other reasons?

Children could use reference material and perhaps consider the types of publicity which may be effective in helping promote a conservation message. The fact that some animals have become extinct in recent times could encourage some investigation and study into the reasons why.

The children may decide to produce a list of animals and plants currently in danger and link with geography in locating where these animals live. Children should consider alternatives to animal products for everyday use. For example synthetic materials. Do we really need these animal products?

GARDEN SNAILS

RESOURCES A collection of garden snails

Children may be able to identify areas around school or in the neighbourhood where snails can be found.

Snails should be kept in a moist environment with a plentiful supply of fresh green plant material. Care should be encouraged if the children wish to handle the animals and a routine for ensuring the safe return of any animal to its original habitat after a short period of study should be policy throughout the school.

Practical Sketching will help the children identify the similarities and differences between individuals in the snail collection. Colour, size, and shape could be recorded. The children may also notice other features or behavioural patterns which they would like to monitor.

Reference books will help encourage the use of correct terminology for the different parts of a snail's body, and collecting information about snails to present to friends could develop a deeper curiosity about these animals, together with a level of study appropriate to a child's interest and ability.

– How do snails travel on different surfaces?
– How do snails protect themselves?
– What do snails prefer to eat?

LEAVES

RESOURCES A variety of leaves collected by the children

Class or group discussion The children could be asked to look closely and make sketches of the leaves in the collection. A discussion could be based around the differences that they notice between the leaves. For example, size, shape, colour, texture and structure.

The similarities and differences in vein pattern, leaf shape and the measurement of size and area of a number of leaves could prove useful for promoting skills of observation.

The leaves could be sorted into groups or sets according to identifiable features such as shape, vein structure, size and surface texture.

From what type of tree do each of the different leaves come? The use of reference material to identify the names of local tree species and the knowledge held by the children should be encouraged.

Information Technology

The computer program *Branch* may help the children devise questions which enable classification of observable features.

A data base built up of the trees in the area could be used to teach the children the name of each tree type, and records could be made of individual tree sizes and features.

ASSESSMENT

Activities 2 to 6 combined offer opportunities to assess:

2.2b Be able to sort familiar living things into broad groups according to easily observable features.

2.4b Be able to assign plants and animals to their major groups using key and observable features.

INVESTIGATION   Variation and the mechanisms of inheritance and evolution

"Are all worms the same? If not, what might have influenced why they are different?"

Starting point
Collect worms from around the school field.

Observing and asking questions
The children should be encouraged to collect worms carefully from different places around the school. It would be interesting to count how many worms were found in each place around the school and for the children to compare their results and give reasons why they think more worms are found in some places than others. The children should be encouraged to observe the worms carefully – how they move, their colour – and to measure their length. Drawings could be made of the worms' distinguishing features and children could research information about the various parts of worms, how they live, reproduce and what they need in order to stay alive and healthy. The children should at all times be encouraged to handle living things with care and consideration and return them as soon as their investigations are over to their correct environments.

Children may ask
Do worms all grow at the same rate?
Does the soil a worm is found in cause them to have different colours?

Predicting and hypothesising
Before the children carry out an investigation to try and answer their question, encourage them to make a prediction about what they think is going to happen and why. For example, children investigating worm growth may predict that all the worms they are investigating will grow at different

Resources
- old aquarium
- soil
- leaves and other plant materials
- seed trays
- rulers and tape measures
- wormeries made in old sweet jars

rates because that is what happens in humans. The children investigating worm colour may predict that this will be influenced by the type of soil in which the worms live. An advanced prediction may involve the children in relating the fact that the parents of a worm will affect colour and speed of growth because of characteristics passed on through genes. (This would be very hard to investigate but may arise in a prediction or in a conclusion the children draw as a result of their investigations.)

Designing and planning the investigation

The children need to collect some worms and to create suitable environments in which the worms can be observed easily and well looked after. The children will need to use their knowledge of what conditions worms need before setting up their investigations. The 'classic' wormery may not provide a suitable environment as it will be difficult to extract the worms for observation.

Investigating length The children should try to pick worms that are approximately equal in length and the same colour. The same food should be fed to the worms and they should be kept in the same environment. The children will need to decide how often they will remove the worms to be measured and how long the investigation should run in order to obtain valid results. Another important decision to be made is how many worms will need to be investigated to ensure a meaningful result.

A problem will also be identifying individual worms in order to plot their growth. This could be overcome by housing each worm separately.

Investigating soil and its effect on worm colour This investigation may require children to collect worms from home and in the school grounds. Worms should be collected from different places with samples of the soil in which they were found. The colours of the worms and the soils in which they were found could be compared. An extension of this could be to put worms found in one soil into another type of soil and to observe if worm colour changes over a period of time. To ensure valid results some worms would need to be kept in the original soil in which they were found.

Recording

The way in which the children record their findings would depend on the investigation being carried out. Graphs and tables would be the best way of recording the findings in the length investigation, but observational drawings and colour-change charts would be most suitable in the colour investigation.

Drawing conclusions

This aspect of the investigation could be very interesting. The children should be encouraged to give good reasons why they think things have happened. If the worms have increased in length by different amounts, what could be the reason? Could it be that some worms have eaten more? Were some worms unhappy in their new 'home' and therefore did not grow, or was it because their parents were larger and they inherited their size from their parents? The colour investigation will give more clear-cut findings. The children may notice that worm colour does vary according to the type of soil in which the worms are found. Does worm colour change when the worms are put in a different soil? If this is the case then it will almost certainly rule out the possibility that colour is inherited from the parents.

Attainment Target 2 Life and living processes

Key Stage 1 strand (iii)
Populations and human influences within ecosystems

PROGRAMME OF STUDY

A Local habitats

Pupils should study plants and animals in a variety of local habitats, for example, playing field, garden and pond.

B Human activity producing change in the environment

They should discuss how human activity produces local changes in their environment.

STATEMENTS OF ATTAINMENT

2.2c Know that different kinds of living things are found in different localities (**A**).

2.3b Know that human activity may produce changes in the environment that can affect plants and animals (**B**).

ACTIVITIES Populations and human influences within ecosystems

MINIBEAST HABITATS

RESOURCES A collection of minibeasts from school grounds

Practical Observe the creatures in the collection through magnifying glasses. Make sketches of them, recording carefully their shape and colour. Look closely at how their bodies are put together, and how they move.

Carry out some research work to find out more information about the creatures that have been observed and why they live where they do.

PLANT HABITATS

Starting point Walk outside school and identify different plants growing in different places.

Children are encouraged to explore different habitats around the school, for example, walls, trees and playgrounds. They could note and try to identify different plants growing in the habitats they explore. The children could draw or photograph plants they found. If at first the children cannot identify the plants, suggest they give them a name which they think suits them. Once back in the classroom, identify missing names and produce reference books about the plants found. Encourage the children to question why the plants are living there.

3 THE BIRD TABLE

RESOURCES A bird table outside the classroom

Observe the birds visiting the table over a period of time. Which birds seem to visit most regularly?

The children should keep a record of the birds visiting the table. Some children could put together a booklet of birds – to help identify the birds seen on the table.

Why do the children think that some types of birds are more regular visitors?

Try observing birds on bird tables in other locations. For example at home and at other schools. Are the results different to your school? If they are – why do the children think this is so?

4 THE WILD AREA

RESOURCES A wild area in the school grounds

Practical Designate an area at school 'wild'. Ensure that the area is notified to the groundspeople – once established it will provide a great deal of interest in comparing the area to mown and weeded parts of the grounds. What are the differences? Which living things can be found on the wild area that cannot be found in other parts of the school? Wild flowers could be sown. Get an 'expert' to come into school to offer advice on suitable wild flowers to sow.

5 CHANGES IN OUR ENVIRONMENT

RESOURCES A collection of photographs of a piece of land, in the country or a city, taken over a period of time

Practical The children are encouraged to identify any differences in the photographs that they have noticed. They should try and give reasons for why these changes have occurred.

Wherever possible the children should be encouraged to consider the effects that the changes have had on the wildlife in the area under study. Children should express how they feel about the changes that have been made. These discussions and observations are particularly relevant in city areas and also where hedgerows have been removed in agricultural parts of the countryside.

Encourage children to collect information on changes made to the environment in other parts of the world.

STUDYING LOCAL HABITATS

RESOURCES A tree or a length of hedge or an area of wasteland

Children discover what creatures live in these habitats – through firsthand experience supported by research.

What effects would the removal of these habitats have on the wildlife in that area?

Children will be amazed at the amount of wildlife that can be supported by a mature tree or a length of hedge.

What happens to the wildlife once the tree or hedge is removed?

A POND DIP

RESOURCES Suitable equipment for carrying out a pond dip

The teacher will need to show the children how to carry out a pond dip and to explain that it can endanger wildlife in the pond if performed incorrectly. Ponds should not be dipped when animals are hibernating.

Careful supervision of all children during a pond dip is important. Suitable clothing is required, and careful plans should be made before the visit. A preliminary visit by the teacher is a good idea to find out what might be found, possible dangers and where children can make safe dips.

Children can find out about the wildlife that lives around the pond as well.

Look at the plants as well as the creatures living in, on and around the pond. Dip into the pond to find any living creatures. Is there a pattern as to where the creatures live – position in the pond – shallow, deep, middle, edge, shady, open? Observe the animals that are found closely. How do they move?

Look for signs of human effects on the pond. What could be done about them and how does it affect the living things that are there?

When the children return to the classroom they can use reference material to make displays of the things they have found. The children could also make large models of the animals they have found based upon their drawings and sketches.

In all these activities the teacher should aim to develop the children's understanding and knowledge of how human activity affects habitats. This should be an underlying message in activities 1 – 7.

ASSESSMENT Activities 1 – 7 will develop the children's understanding of:
2.2c Know that different kinds of living things are found in different localities.
2.3b Know that human activity may produce changes in the environment that can affect plants and animals.

INVESTIGATION Populations and human influences within ecosystems

"What conditions do minibeasts like to live in?

Starting point
A walk around the school grounds, or a wild area to study minibeasts.

Observing and asking questions
The teacher needs to discuss with the children safe and careful handling of any living creatures that they find. Careful thought needs to be given as to how the minibeasts that are found are going to be kept. Children need to be encouraged to use a variety of techniques for recording what they find – drawings, tables, photographs, tape and video recordings.

Ask the children to note if certain minibeasts live in similar types of conditions. Does there seem to be a pattern?

Encourage the children to look carefully at the minibeasts and to describe their appearance and how they move. The children should be encouraged to sketch the areas where the minibeasts were found and to make lists or descriptions of the habitat in which they were discovered. The minibeasts can be collected using a pooter, as shown in Figure 1.

Children may ask
Do minibeasts choose to live where they were found because they like that type of home?
Do different minibeasts prefer the same type of living conditions?

Encourage the children to identify one thing to investigate.

Resources
- magnifying glasses, microscopes, hand lenses
- pooters
- observation dishes, suitable containers for holding specimens
- shoe boxes
- old fish tank
- natural material
- drawing, painting, sketching equipment
- camera or video camera
- tape recorder

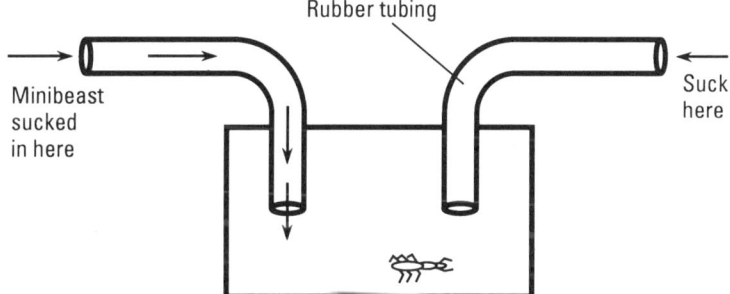

Figure 1 A pooter

Predicting and hypothesising
Before the children carry out the investigation of their choice encourage them to make a prediction about what they think is going to happen and why.

Designing and planning the investigation
The children could investigate two different types of habitat. One that is the same as that where the minibeasts were found, and the other dissimilar. The children should predict what they think will happen.

Children need to decide what things they are going to provide in each of the two habitats. Decisions will be based upon the children's observations when they were out on the school field. Decisions also need to be made about how to set up the habitats so that the minibeasts can pass freely between the two.

How long will the minibeasts be left to make a choice and how can the children ensure that they are treated carefully?

Children need to decide how many times they will carry out the investigation and how they will record their results.

Recording

Children should devise ways of recording their findings and displaying them for others to see. The results could be recorded in illustrated form along with an account of how the investigation was carried out. Emphasis should be placed on the children explaining how they ensured the safe handling of the minibeasts and how they made the investigation fair.

Wherever possible use IT to assist in the recording and presentation of work. A video camera would be an ideal means of recording and displaying to others how the minibeasts reacted to choosing a habitat.

Drawing conclusions

What did the children find out? Did the minibeasts (for example, woodlice) choose the environment that the children thought they would? Encourage the children wherever possible to think of reasons for what happened.

Do other animals have preferences for where they live? If so, find out about the animals' preferences and why, using reference material.

All living things should be returned, as soon as possible, to their natural environment on completion of the investigation.

Attainment Target 2 Life and living processes

Key Stage 2 strand (iii)
Populations and human influences within ecosystems

PROGRAMME OF STUDY

A Animals, plants and their habitats

Pupils should explore and investigate at least two different habitats and the animals and plants that live there. They should find out how animals are suited to these habitats and how they are influenced by environmental conditions, including seasonal and daily changes, and measure these changes using a variety of instruments.

B Caring for living things

They should develop an awareness and understanding of the necessity for sensitive collection and care of living things used as the subject in any study of the environment.

C The effects of human activity on the environment

They should study aspects of the local environment affected by human activity, for example, farming, industry, mining or quarrying and consider the benefits and detrimental effects of these activities.

D Living things (competition and pollution)

They should be made aware of the competition between living things and their need for food, shelter and a place to reproduce. They should study the effects of pollution on the survival of living things.

STATEMENTS OF ATTAINMENT

2.2c Know that different kinds of living things are found in different localities (**A** and **D**).

2.3b Know that human activity may produce changes in the environment that can affect plants and animals (**B**).

2.4c Understand that the survival of plants and animals in an environment depends on successful competition for scarce resources (**D**).

2.5c Know how pollution can affect the survival of organisms (**D**).

ACTIVITIES Populations and human influences within ecosystems

CREATING A DATA BASE

RESOURCE A suitable data base for recording information about animals observed

Information Technology If you have an Archimedes computer it is recommended that you use *Junior*

Pinpoint by Longman Logotron to create and interrogate your data base.

Decide whether the teacher or the children are to set up the data base. This will depend upon their previous experience and the level at which they are working.

Gathering information

What type of information have you collected about the animals you are studying? How might you group the information – for example, habitat, how they move, length, colour? You may wish to decide on the headings for your data base before you observe the animals to help you look for the right things.

Sorting your information

Create suitable headings on the data base. For example, can the creature fly, what is its name, how long is it, what does it eat, does it prefer the dark, what colour is it, where is it found?

Interpreting your information

What questions might you ask that the data base will give you information about? Print out some of the information as a graph. What do the graphs tell you about the creatures? List all the creatures that can fly. Why has the computer only found a few records to the question you have asked?

Interrogate the data base

Children could ask the computer to give information related to a number of questions that the teacher asks them. *I want to know all the creatures that fly and are longer than 15 mm.*

Can others who were not involved in drawing up your data base find the answers to your questions?

CREATING AN ENVIRONMENT

RESOURCES Woodlice or other minibeasts found in the local environment

Observation

Look under stones, bark, leaves, hedges, soil, walls etc. for a variety of creatures. Look particularly for woodlice, worms, spiders, ants, snails and beetles.

Remember to always place things back exactly how you found them. Do not destroy another creature's environment. Encourage the children to always treat creatures and their environments with care.

Do all the creatures live in the same place in the same type of environment? Look closely at their environment. What is it like? Could you re-create that environment in the classroom in order to study the creatures more closely? Observe the creatures closely in their habitat. How do they move? What special features can you see? How do you think these creatures see, hear, smell and feel? How might we find out more about these creatures?

Take careful note of all the details that will help you re-create the right environment in which to study the creatures back in the classroom. The environment may take some time to prepare.

WASTE LAND

RESOURCES An area of waste land

Practical

Make a study of waste land throughout the school year. Make contact with the owner and find out as much as you can about how it has been used in the past. Has it always been waste land? Can you find evidence of what it was previously used for? Can you obtain any old photographs of the land – your local museum or paper may have photographs they may let you copy or borrow.

Decide what you might observe throughout the year or term. You could make a survey of several different areas or mark off one small section for study. How might you record how the land's appearance changes over time? Could you keep a photographic record?

Decide when you are going to study the waste land. You may decide to observe the changes at the same time of day once a month. Will what you observe vary depending upon the time of day you visit?

Study the main plants growing on the land. Do certain plants only grow in certain areas? Why? Can you find out why they only grow in these areas? How do the plants spread from one area to another?

Cross-curricular links
Geography

Can you make a map of the waste land showing all its main features and those which you have decided to observe? Make sure you identify the main physical features on your map so that others can use it to identify the location of the area you are studying. How has your map changed by the end of the year? Record these changes and discuss how the changes affect the plants and animals that live in, or use the waste area.

4 A FIELD

RESOURCES

A field situated nearby that can easily be observed over a period of time

Practical

Make a study of the field throughout the school year. Make contact with the owner and find out as much as you can about how it has been used in the past. Has it always been used in the same way, for example to grow crops or for grazing?

Decide what you might observe throughout the year or term. How might you record how the field changes over time? Could you keep a photographic record?

Could you arrange for the farmer or owner to come and talk to the children about what happens to the field? What questions might you and the children ask your visitor?

How does the farmer prepare the field? What are the perfect conditions for the crop to be grown in the field? Could you ask for some seeds and see if you could grow the crop in the classroom? How might you record its growth? You could compare its growth with the farmer's crop. Are there any differences in the way the crops grow? What might have caused the differences?

Does the farmer harvest all of the crop? Is any of it recycled, burned or ploughed into the land? Ask the farmer for a sample of the crop that you can bury and observe how it changes over a period of time.

Cross-curricular links
Geography

If possible plan to observe two differing localities over time so that comparisons can be made.

5 SURVIVAL

Practical

Research into why the Giant Panda is nearing extinction.

Find out about where it lives naturally and what it eats. Write to zoos and the World Wide Fund for Nature.

Find out about other endangered species and how man has had a part to play in their dwindling numbers. Encourage children to identify the reasons why certain animals are fighting for survival.

6 WHAT LIVING THINGS NEED IN ORDER TO SURVIVE

Class or group discussion

What do all living things need to survive?

Practical

List things that the children feel all living things need in order to survive. Children should develop an understanding that all living things require food, shelter and a place to reproduce.

Draw up a list of those animals that have become extinct, those that are threatened with extinction and those that are thriving. Encourage them to gather information on how the animals in each group survive in terms of having shelter, food and somewhere to reproduce.

The children can then explore what effects upon food, shelter and reproduction have caused each animal to survive, become extinct or near extinction.

Use examples such as the Dodo to explore the ways in which animals can become extinct.

Explore ways in which man can help in the survival of plants and animals in their local environment and throughout the world.

7 CLASSIFYING PLANTS AND ANIMALS

RESOURCES

Make a collection of photographs, drawings and pictures of plants or animals. Ask the children to think of ways in which they might group the different plants or animals

Practical

Explain that scientists have ways in which they group different plants or animals. Explain the ways in which plants and animals might be grouped. Develop an understanding of what makes a plant a member of one of these groups.

Children should be encouraged to consider classifying plants into groups such as ferns, conifers and flowering plants.

Animals could be grouped into vertebrate groups.

Give children the opportunity to classify animals and plants by observing closely their different features and using keys.

ASSESSMENT   This activity does not provide any direct assessment opportunity but contributes towards the development of understanding of the Programme of Study for this strand.

INVESTIGATION   Populations and human influences within ecosystems

"How do we clean oil from birds' feathers?

Starting point
A video, pictures or newspaper reports into how oil pollution is affecting birdlife. (For example, the Shetland disaster, Alaska oil disaster and Kuwait after the Gulf War.)

Observing and asking questions
Children could be involved in discussion and research work about oil and the oil disasters outlined above. Questions that could be researched by the children that would promote discussion and develop knowledge could centre around why the oil is so dangerous to birdlife and wildlife generally, and how the birds we help recover from the effects of oil pollution. The children could be encouraged to write to oil companies to discover more about oil and its production and what is being done to reduce the chances of disasters such as those listed above occurring again.

Children may ask
Which cleaning agent will be best at removing oil from feathers?
Does the quantity of cleaning agent affect how easily oil is removed from feathers?
Does it matter how concentrated the cleaning agent is?

Predicting and hypothesising
Before the children carry out their investigation encourage them to write down or tell you what they think will happen and to give reasons why. There will be a need to undertake a range of activities to develop children's understanding of how cleaning agents work when breaking down the oil, so that children can make informed predictions.

Designing and planning the investigation
If children set up an investigation in which they choose appropriate cleaning agents and investigate which is most effective for cleaning the oil they will need to decide which cleaning agents they are going to investigate. They also need to make sure that all the other aspects of the investigation are kept constant. For example, they must ensure that any feathers used are the same and that they have the same quantity of oil on them. The same amount of water must be used and the method of cleaning, rinsing etc. must remain constant.

The setting up of the investigation must be carefully considered for safety reasons. Children should work in an area where any spills will not affect other children. They should wear gloves throughout the investigation.

Decisions need to be made as to who will do what within the group. Children also need to decide what is meant by effective cleaning – will it be simply that all the oil has been removed, or something more; for example, the texture of the feather after cleaning?

Recording
Careful thought will need to be given as to how the results of the investigation are going to be recorded. The children may decide to mount the feathers

Resources
- oil
- feathers
- liquid soap (different types)
- bar soap
- cloths and covers for tables
- measuring cylinders
- water
- old bowl

after they have been 'cleaned' and dried so that other children can judge which cleaning agent was most successful. It may well be necessary to keep a feather that has not been cleaned and a feather which has not been 'polluted' in order to make valid comparisons.

Drawing conclusions

The children draw conclusions from their investigations and relate back to their original prediction. Perhaps the children could put the cleaners into their order of cleaning efficiency.

Another question that may develop from this investigation might be how much liquid cleaner gives best results when used to clean oily feathers? Is there an optimum amount?

Within the same context children could investigate the use of cleaning agents on breaking up oil on the surface of water.

As follow-up work the children could contact and inform the RSPB about their findings and request information about how the organisation cleans polluted birds.

Attainment Target 2 Life and living processes

Key Stage 1 strand (iv)
Energy flows and cycles of matter within ecosystems

PROGRAMME OF STUDY

A Studying living things in the local environment

Drawing upon their study of living things in school and the local environment, they should be introduced to the idea that plants are the ultimate source of all food in the living world.

B Waste, decay and improving the environment

They should investigate how far everyday waste products, for example, garden refuse, paper, plastic materials and cans decay naturally and use this knowledge to improve the appearance of their local environment.

STATEMENTS OF ATTAINMENT

2.2d Know that some waste materials decay naturally but do so over different periods of time (**B**).

2.3c Know that green plants need light to stay alive and healthy (**A**).

ACTIVITIES Energy flows and cycles of matter within ecosystems

 DECAY

RESOURCES Children's lunchboxes

Practical A collection of waste products from lunchtime could be sorted into groups. For example, wrappings, silver foil, cans, plastic containers, food.

The children should identify the different types of materials that are thrown away after any school lunchtime.

Why are these things thrown away?
What happens to them?
Are there any items which could be saved and used for another purpose?

The children need to investigate what happens to everyday waste products when they are left to decay naturally.

One collection could be buried and other items could be left to decay in the air. Comparing what happens when they are left over a period of time could include leaving some waste indoors and other materials outside.

Health and safety Decaying food should not be touched by the children or left in places where it is accessible around the school.

A diary to show the rate of decay and the changes could be kept by the children.

Which materials decay quickly?
Which materials fail to decay?

Which waste materials seem to pose special problems if they are carelessly left or forgotten about?

2 THE LOCAL PARK

RESOURCES | A visit to the local park or an area of waste land

Practical | On your visit to the local park (or waste land) help the children to identify which things have been placed in the area by humans and which are natural. The children need to have a view of which aspects of the park are useful and pleasing and which features are ugly or harmful. Bring the children's attention to litter that may be present and relate to work on decay in Activity 1.

How is the park kept tidy?
What changes happen to the park in summer or winter?

Look at the litter left lying in the park and take note of the different materials. Some of these materials may be the same as the ones used in Activity 1. Relate the time taken for the different materials to decay and the effects of the litter on the environment. Think of the effect of the litter on wildlife and people that visit the site. Children's views for improving and caring for this local area should be discussed.

3 APPLE CORES

RESOURCES | A selection of apple cores

Practical | Apple cores could be placed in a number of areas around school in order to find out the conditions which seem to speed or delay the process of decay.

Try keeping an apple core: in an airtight container within a refrigerator, in a moist area outside, on a radiator, buried in soil.
Help the children to realise that temperature, air and water each have a role in the process of decay.

Can the children suggest another cause of decay? For example, microbes.

ASSESSMENT | Activities 1- 3 provide opportunities to assess:
2.2d Know that some waste materials decay naturally but do so over different periods of time.

4 CARROT TOPS

RESOURCES | Carrot tops

Practical | Carrot tops grow leaves very quickly when given light and water.

The children would enjoy creating a 'mini-forest' from carrot tops and could investigate:

– the rates of growth of different carrot tops;
– the volume of water needed to sustain each growing plant;
– the best location in the classroom;
– what happens if a plant is deprived of water or light.

Some of the carrot tops could be exposed to light for limited periods and their growth rates compared with the growth rates of tops receiving continuous daylight. It may be possible for the children to investigate what happens to tops grown in continuous artificial light.

Cress seeds and mung bean seeds are also suitable plants for this activity as they grow rapidly.

ASSESSMENT

Activity 4 offers an opportunity to assess:
2.3c Know that green plants need light to stay alive and healthy.

INVESTIGATION

"What do you think happens to waste materials when they are buried?"

Energy flows and cycles of matter within ecosystems

Starting point
A collection of waste materials collected after lunchtime from children's lunch boxes.

Observing and asking questions
Children should be encouraged to sort and classify the waste materials, once they have been washed and cleaned, handling them with plastic disposable gloves. The teacher should ensure that there are no waste materials that could harm the children, for example sharp tins. Observational drawings could be made of the waste products. Discussion could centre around what happens to the waste generated in school and where the children think it goes. A large amount of work can be undertaken researching how much waste is generated at school and at home and children can estimate how much waste is produced by the school in a term or a school year. Other work that could be undertaken by the children could centre around waste disposal. The more able children could find out about what the products originated from, and how they were made. The children should be encouraged to identify similarities and differences between the waste products. For example those which are natural and those which are man-made. The teacher could ask the children what happens to waste products when they are buried.

Children may ask
Does burying speed up the rate of decay?
Do different materials decay at different rates when they are buried?
Does the depth at which materials are buried affect how quickly they decay?

Predicting and hypothesising
Before the children carry out their investigation they need to be encouraged to give a prediction about what they think will happen and give a reason why they think this is so.

A prediction could take the form of: *I think that the things that we bury will decay quicker the deeper they are, because the weight of the soil will be greater and that will help them to decay.* Another example could be: *The things that are buried will decay more quickly than those that are not because it's warm and damp underground and that helps things to decay.* Children may also make predictions based upon their knowledge of materials, for example: *I think that the apple core and orange peel will decay more quickly than the yoghurt pots and cling film because they are natural and are made to decay whereas the man-made things are not made to decay. Man-made things decay slowly or not at all.* Predictions such as these will lead to many interesting questions, discussion and research work once the investigation has been completed. If the teacher puts in some biodegradable materials then the results may be different to the

Resources
- spades
- variety of waste materials (natural and man-made; some biodegradable)
- metre rules
- plastic disposable gloves
- video camera (not essential)
- camera (not essential)

children's predictions and will lead to further work on what biodegradable products are and why they are so important to the environment.

Designing and planning the investigation

When the children have decided which question they would like to investigate, encourage them to decide which variable they will investigate (the thing they are going to change in each investigation) and the variables they will keep constant.

Burying In this investigation the children simply decide which waste materials they are going to use; bury some of them and keep the remainder above ground. In order for the test to be fair and to obtain comparable data, it is important that those things which are buried are the same as those things which are kept on the surface. The children need to decide how deep to bury the waste materials, how often to inspect the materials and how long to run the investigation before drawing their conclusions.

Different materials In this investigation the children will need to decide how deep to bury the materials and which materials they are going to use. In order to obtain substantially different results encourage the children to choose both natural and man-made materials. Each material could be buried in a separate hole to the same depth. The children need to decide how long to leave the materials to decay.

Depth of burying The children will need to select at least two or three different materials to bury in each hole. The depth of each hole should vary – but the materials in each hole should be exactly the same. The children need to decide how long to leave the materials before digging them up.

Other factors that the children will need to consider are the size of the materials put in to each hole. They will need to be the same so that direct comparisons can be made. The dug holes will need to be the same size and dug in similar places, so that variables such as moisture and soil type are comparable. The positions of the holes will need to be marked clearly so that they can be found without difficulty.

Health and safety Ensure that the children are always wearing gloves when handling decaying materials and that they are disposed of by the teacher as soon as possible once the investigation is complete.

Recording

The most obvious way of recording the findings of the investigations will be through direct observation and description of the waste materials, both before they are buried and then at different times during the investigation and finally at the end of the allotted period of time. Children's descriptions should be sufficient to indicate the level and amount of decay. More able children may be encouraged to measure length, weight or area of the waste materials before and after the investigation, which would give more quantifiable results. Photographs or video recordings of the investigation and the children's descriptions would be an even better way of recording results.

Drawing conclusions

What did the children see happen to the waste materials? The children's attention should be drawn to anything unusual that they had not predicted that happened to the waste materials. Every effort should be made to encourage the children to give reasons for what happened. This could be followed up by research work at a simple level. The children should be encouraged to relate what they have found out to everyday situations. What have we found out from our investigation that tells us about the importance of care when disposing of waste materials and the effect upon the environment in which we live?

Attainment Target 2 Life and living processes

Key Stage 2 strand (iv)
Energy flows and cycles of matter within ecosystems

PROGRAMME OF STUDY

A The Sun as a source of energy and food chains

Pupils should be introduced to the idea that green plants use energy from the Sun to produce food and that food chains are a way of representing feeding relationships.

B Decay and waste disposal

They should investigate the key factors in the process of decay such as temperature, moisture, air and the role of microbes. They should build on their investigations of decay and consider the significant features of waste disposal procedures, for example in sewage disposal and composting, and the usefulness of any products.

STATEMENTS OF ATTAINMENT

2.2d Know that some waste materials decay naturally but do so over different periods of time (**B**).

2.3c Know that green plants need light to stay alive and healthy (**A**).

2.4d Understand food chains as a way of representing feeding relationships in an ecosystem (**A**).

2.5d Know about the key factors in the process of decay (**B**).

ACTIVITIES Energy flows and cycles of matter within ecosystems

Teacher's note Apart from Statement of Attainment **2.4d**, this particular strand lends itself to being covered and assessed by investigation. It is suggested that teachers undertake the following two investigations to ensure coverage and assessment of this strand:

AT2(i) Investigation (page 35)
This will provide opportunity to assess:

ASSESSMENT 2.2a Know that plants and animals need certain conditions to sustain life.
2.3c Know that green plants need light to stay alive and healthy.

The investigation must be extended to include a focus upon light rather than soil if assessment against **2.3c** is to be made. Within this investigation children should be developing an understanding that plants need light in order to stay alive and healthy.

AT2(iv) Investigation (page 65)

In order to ensure coverage of the remainder of this strand teachers should undertake the following activities.

DECAY OF NATURAL MATERIALS

RESOURCES

A number of natural materials placed in an area of the school where they will be moved and can be observed over a period of time

Class or group discussion

What is happening to these materials?

Compare the colour, shape, texture, size and weight of decaying materials with similar items in good condition.

Are there any creatures living on the decaying materials? Introduce the children to microbes and their role in the decay of materials.

Children need to be introduced to the key factors in the process of decay. These are temperature, moisture, air and microbes.

How might decay be prevented?

Practical

Place a number of natural materials outside in an area where they can be observed over a period of time. Set up a recording system so that groups of children can keep records of the effects of the elements of each object. Photographs and detailed sketches will be invaluable in recording the rate and type of decay.

ASSESSMENT

This activity will add to the children's understanding of:
2.5d Know about the key factors in the process of decay.
This activity should be undertaken prior to the investigation for the strand.

FOOD CHAINS

RESOURCES

A collection of charts, drawings and reference materials on the subject of food chains

Teacher demonstration

The teacher could introduce a simple food chain using animals that children will be familiar with, and develop the understanding that:

– Animals feed on one or all of: organisms, other animals and plants.
– Plants are producers: producing their own food and providing consumer's food.
– Animals are consumers: feeding on food produced by plants.
– Food chains usually begin at the producer level.

Practical

Once the children are introduced to the idea of food chains, they enjoy drawing their own. The children could use reference books to compile different food chains.

Which is the longest food chain that you can research?
Which groups of animals are seldom absent in many food chains?
Explore how the children's food chains can be destroyed. Discuss the effect of taking out one member in each of the food chains.
Look for examples where food chains have been broken that children have read about in books, or that have been in the news recently.

Do all the food chains confirm the Sun as the starting point for all feeding relationships?

Activities 1 and 2 combined provide opportunities to assess:
2.4d Understand food chains as a way of representing feeding relationships in an ecosystem.

INVESTIGATION

"What affects how a compost heap decays?"

Resources
- plastic bags
- grass cuttings
- water
- thermometers
- scales
- plastic sweet jars

Energy flows and cycles of matter within ecosystems

Starting point
A discussion about why compost heaps are created and their use to gardeners.

Observing and asking questions
This part of the investigation could centre around discussion of compost heaps and why they are created and used. The children should be encouraged to find out all they can about compost heaps, perhaps asking local gardeners how they make their compost heaps, how long they leave them for and how they use the compost.

Children may ask
Does the moisture content of a compost heap affect how well it decays?
Does the amount, or weight, of a compost heap affect how well it decays?
Does a compost heap need air to help it decay?
Does the outside temperature affect how well a compost heap decays?

Predicting and hypothesising
Before the children carry out the investigation they should always try to predict what they think will happen and give reasons why. If the children have undertaken work previously on the area of decay and its causes, they are likely to be able to give informed predictions. For example: *The compost that is moist will decay more quickly because micro-organisms that speed up decay prefer moisture and grow quicker in these conditions.* Likewise, a knowledgeable child may predict: *The compost heap with less air will take longer to decay because micro-organisms that are in the air will be unable to 'land' on the compost and speed up the decay.*

Designing and planning the investigation
The children should be encouraged to make their investigations as fair as possible by investigating one variable at a time. They must also ensure that each compost heap is kept in similar conditions so that outside factors do not affect the final results of their investigation. Every care should be taken with this type of investigation as children should not be exposed to rotting materials. Children should wear gloves at all times when handling the materials.

Moisture content In this investigation the children will need to ensure that their 'compost heaps' are made of the same materials and are the same mass. The variable being investigated is moisture content so every effort should be made to make sure that each 'compost heap' has an equal moisture content at the start of the investigation. This could be done by using grass cuttings and then allowing them to dry for several days. Then the children should put the same mass of grass cuttings into several sweet jars. The children then need to decide how much moisture they will give to each 'compost heap', making sure, of course, that one does not receive any, thus acting as a control. The children examine the grass cuttings over a period of time to see what has happened.

Mass of a compost heap The children simply need to change the mass of each of the 'compost heaps'. They too could also use grass cuttings in sweet jars, varying the amount of grass they put into each jar.

Air content in a compost heap The children could put an equal quantity of grass cuttings into two identical plastic bags. All the air could be squeezed out of one of the bags and the bag sealed. The other plastic bag could be sealed with air inside and also have air holes pierced in it. The children would need to decide how long the samples will be left before comparing results.

Effect of ambient temperature on a compost heap The following variables should be kept constant: the type of material that makes up the 'compost heap', the mass of the material in each 'compost heap', the time period of the investigation. The ambient temperature for each 'compost heap' should be varied. Keeping ambient temperature relatively constant is very difficult and it will probably be easier to work with one 'compost heap' in a warm environment and one in a cold environment, rather than trying to keep the temperature exact. If the teacher has the facility in their school to keep the samples at an exact temperature then this will, of course provide a much fairer investigation.

Before the children record any of their findings and begin to draw conclusions they will need to decide what they understand to be 'good' decay. The temperature of the heap increases during the decaying process. The children could be encouraged to record temperatures of the various 'compost heaps' as an indication of how well the process of decay is occurring. The mass of each of the 'compost heaps' could also be recorded. But probably the most effective form of judgement will be the children's direct observations of the decaying heaps.

Recording
Depending on how the children have decided what constitutes 'good' decay, this will have an influence on what and how they record. Direct observations could be given with drawings and descriptions of what has happened. If temperatures and masses have been recorded these will lend themselves to quantitative graphs, tables and charts.

Drawing conclusions
The children should relate their conclusions to their findings and if data have been collected, make reference to these data in their description of what happened.

The children could be encouraged to relate what they have found out to real life situations. Passing on their information to local gardeners, so as to offer advice on how to set up a good compost heap, for example.

 Other interesting conclusions that could be drawn and could lead to further investigation are such things as the role of microbes in compost heaps. Are there more microbes on compost heaps that are decaying quickly, compared to heaps that are decaying slowly? Local secondary schools may lend petri dishes with nutrient agar on which moulds taken from different compost heaps at different stages of decay can be grown. (This will need to be closely supervised as children should not be allowed to come in contact with moulds.)

Attainment Target 3 Materials and their properties

Key Stage 1 strand (i)
The properties, classification and structure of materials

PROGRAMME OF STUDY

A The properties of materials

Pupils should collect and find similarities and differences between a variety of everyday materials. These should include natural and manufactured materials such as rocks, soil, air, water and other liquids, cooking ingredients and metallic objects. They should explore the properties of these materials referring, for example, to their shape, colour and texture, and consider some of their everyday uses. They should see how some can be changed by simple processes such as dissolving, squashing, pouring, bending and twisting.

STATEMENTS OF ATTAINMENT

3.1a Be able to describe the simple properties of materials (**A**).

3.2a Be able to group materials according to observable features (**A**).

3.3a Be able to link the use of common materials to their simple properties (**A**).

ACTIVITIES The properties, classification and structure of materials

SORTING MATERIALS

RESOURCES

Gather together a selection of different materials, natural and man-made, most of which children will be aware of. Add one or two unusual materials, such as natural sponge, cork, pumice (which is a rock that floats), Plasticine and clay (in different forms)

Class discussion

Show the materials to the children asking them if they know what they are and where they came from. Explain carefully the ones that are unfamiliar to the children. Begin to develop a language for describing the different materials in terms of shape, colour, texture and use.

Practical

Children may use texture, shape, colour or where the material comes from to sort the materials. Encourage the children to think of as many adjectives as they can to describe the materials. Encourage the children to sketch and describe the materials so that they can be recognised by others. Young children enjoy seeing older children using their information in order to identify a material. Emphasis should be placed on developing the children's ability to describe different materials.

Information Technology

IT can be used here with *Branch/Sorting Game* to help the sorting. Describe the properties of the materials – try a guessing game, with children describing three properties of a material, then another child picking the material described.

2 SOIL

RESOURCES | A collection of different soils. These can be made up quite easily by gathering a selection of differently sized gravels and sands from a builder's yard and mixing in different proportions with soil

Practical | Children should examine the soils closely (wearing disposable gloves) describing the features of the soils – texture, colour, constituent parts. What can be found in the soil? Are all the soils the same? Children could progress to investigate how the soil changes when it is wet or dry.

Children could investigate how well different soils drain – are some soils better for growing things in than others?

3 THE WATER TRAY

RESOURCES | The 'water tray'. This should contain a variety of objects that squirt and pour

Practical | Try pouring water with a variety of pouring devices – encourage the children to describe what they feel and see. If the children pour the water from different heights, what do they notice?

Using different types of squirters, such as syringes, washing-up liquid bottles, pumps: does the amount of pressure applied affect how the water squirts?

Try mapping the pathway of the water.

Teachers should be developing the children's understanding of the properties of water in terms that it can be poured, squeezed through tiny holes, run along certain objects etc.

ASSESSMENT | Activities 1 to 3 combined provide opportunities to assess:
3.1a Be able to describe the simple properties of materials.
3.2a Be able to group materials according to observable features.

4 KITCHEN UTENSILS

RESOURCES | A collection of kitchen utensils

Class or group discussion  | Children could be asked about why different kitchen utensils are made from different materials. Firsthand experience could be given by placing wooden spoons, and metal spoons on radiators to observe the different rates at which each material heats up. Only the teacher should undertake such an activity. Children could be introduced to the idea that some things that do not look hot could be. The children should be shown the correct way of testing if something is hot.

Relate the function of the different utensils with the materials they are made from.

5 DECAYING FRUIT

RESOURCES | Fresh and decaying fruit

Children could discuss why the fruits are decaying. What is happening to them? Investigate what food wrappings keep fruits fresh for longer. Children could investigate foil, cling film and sandwich bags. Which of the materials is best? Why do the children think this is so?

INSULATING MATERIALS

RESOURCES Fish and chips wrapped in paper

Practical Why do fish and chip shops sometimes wrap your fish and chips in paper? Children could investigate the properties of certain materials at keeping things warm – try different papers and fabrics. Which is best? Is it the one fish and chip shops use? Children could be introduced to the idea of insulation and why it is important.

Teacher demonstration Place some warm water in a number of containers (all the same volume, but different materials) and measure their temperature. Observe which stays warm the longest. This is an opportunity for teachers to introduce children to using a thermometer and to the planning of a fair test. Which materials kept the water warm the longest?

Where else do we use insulation? What types of materials are used for insulation?

 EXPLORING EVERYDAY MATERIALS

RESOURCES A collection of everyday materials – wood, stone, metal, cooking ingredients, liquids and plastics

Practical Children examine the materials closely, describing them using their senses. They then find a range of ways in which they are used in everyday life.

The children could find real examples or find pictures of the things that make use of or are made from the materials on display.

The children should be able to say why the material is suitable for its use.

ASSESSMENT Activities 4 to 7 combined will provide opportunities to assess:
3.3a Be able to link the use of common materials to their simple properties.

 MIXING DRINKS

RESOURCES Selection of drinks made from cordials. (Use different flavourings and different concentrations)

Practical Ask the children to describe the taste of the drinks, and why they taste different.

Talk to the children about other substances that you can add water to. Do these substances behave in the same way? Discuss substances that disappear, those that remain the same, and those that change colour.

Investigate the following:

– equal mass of different substances in the same volume of water;
– different volumes of water with the same mass of same substance;
– different water temperatures (hot, cold, luke warm) and the same mass of same substance;
– different masses of the same substance with the same volume of water.

Does the number of times the mixture is stirred make a difference?

This activity would be very useful to undertake before carrying out the investigation for this strand.

ASSESSMENT   This activity helps to develop children's understanding of the Programme of Study for this strand but has no direct assessment opportunity.

INVESTIGATION

"What will paper towels do?"

Resources
- a variety of paper tissues and towels
- water
- rulers
- stop clocks
- measuring cylinders
- magnifiers

The properties, classification and structure of materials

Starting point
A collection of paper tissues and towels.

Observing and asking questions
Encourage the children to examine the collection of paper towels and tissues. The children should describe how each sample feels and what it looks like – after close observation with magnifiers. Emphasis here should be placed upon using the senses to describe the nature of the paper towels and tissues. Children should be encouraged to explain what they think the towels would be best for. For example, mopping up, drying hands or surfaces.

Children may ask
Which one of the towels is best at mopping up water spills?

Predicting and hypothesising
Encourage the children to make a prediction about what they think will happen and why. For example: *I think the paper towel will be best at mopping up water because we have paper towels at school for drying our hands.*

Designing and planning the investigation
The children should try to ensure that they keep the test fair by keeping everything constant apart from the paper towel or tissue they are testing.

Children could choose a minimum of three paper towels or tissues to test. Make sure the pieces of towel or tissue are all the same size. In order to make the test fair, the children will need to decide how much water will be 'mopped up', over what period of time the experiment will last and what they mean by best at 'mopping up'. Young children find it difficult to ensure that all aspects of the investigation are fair so encourage them to investigate just one variable at a time.

Decisions will need to be made by the children about how they are going to carry out their investigation and how they are going to record their findings.

The children carry out the investigation making sure that they are careful with measuring the sizes of the pieces of paper towel being used and the quantity of water to be mopped up. Consideration needs to be given to the management of this investigation to ensure the minimum of mess.

Recording

Children record their findings and explain what happened. The pieces of towel or tissue could be mounted in order of how well they 'mopped up' water. Children could record their findings as a graph. *Data Sweet* (Archimedes) could be used in this context. Emphasis should be placed on making sure that the children record and present their findings in such a way as to allow others to understand clearly what they have found out.

Drawing conclusions

Were the children's predictions correct and were they able to keep the test fair?

Some children may wish to take their investigations further.

How successful are paper towels for 'mopping up' other liquids such as vegetable oil?

Does the time given for 'mopping up' have any effect? Are some paper towels quicker than others?

Does the size of the paper towel have any effect on the speed of 'mopping up'?

The children could write to the companies who make the tissues and towels and tell them about their investigations and what they found out, requesting information about how the towels are made, and why they are made in the way they are.

Attainment Target 3 Materials and their properties

Key Stage 2 strand (i)
The properties, classification and structure of materials

A Properties of materials, solids, liquids and gases

Pupils should investigate a number of different everyday materials, grouping them according to their characteristics. Properties such as strength, hardness, flexibility, compressibility, mass ('weight'), volume, and solubility should be investigated and related to everyday uses of the materials. Pupils should be given opportunities to compare a range of solids, liquids and gases and recognise the properties which enable classification of materials in this way.

B Acids and alkalis

They should test the acidity and alkalinity of safe, everyday solutions such as lemon juice using indicators which may come from plants such as red cabbage.

C The dangers of household materials

They should know about the dangers associated with the use of some everyday materials including hot oil, bleach, cleaning agents and other household materials.

D Dissolving, evaporating, separating and purifying

Experiments in dissolving and evaporation should lead to developing ideas about solutions and solubility. They should explore ways of separating and purifying mixtures such as muddy water, salty water and ink, by using evaporation, filtration and chromatography.

3.2a Be able to group materials according to observable features (**A**).

3.3a Be able to link the use of common materials to their simple properties (**A** and **C**).

3.4a Be able to classify materials as solids, liquids and gases on the basis of simple properties which relate to their everyday uses (**A**).

3.5a Know how to separate and purify the components of mixtures using physical processes (**D**).

3.5b Be able to classify aqueous solutions as acidic, alkaline or neutral using indicators (**D**).

ACTIVITIES The properties, classification and structure of materials

 BUILDING MATERIALS

RESOURCES A collection of building materials, such as granules for cavity insulation, 'sponge' for pipe-lagging, polythene as used for damp courses, a blue-brick and roofing felt

Class or group discussion Discuss the various materials and their uses. Include some materials that may not be familiar to the children. Refer back to the work undertaken at Key Stage 1 (pages 67-71) and the ways in which the children described materials and their properties.

Practical Sort the materials into groups and describe what their properties are. Why do these properties make them particularly suitable for the job they do? Children could find out if these materials have always been used for the job they now do.

If not, what was used in the past, and why is it no longer used?
What other materials might be used for the same task?

 WOOD

RESOURCES A collection of different types of wood

Class or group discussion Discuss the different types of wood and where they came from. Develop the vocabulary used at Key Stage 1 (pages 67-71) for describing the properties of different materials.

Practical Many investigations can be undertaken into the properties of wood.

They include flexibility and hardness. Do they all float in the same way, do different woods have different speeds of decay?

What is wood used for around the world?

Children could research into why certain types of wood are used for certain purposes. Reference should be made as to why the properties of particular types of wood make it most suitable for the job it does.

Time should be given to discuss the importance of preservation of our trees and why it is so important.

ASSESSMENT Activities 1 and 2 combined with activities 4 to 7 from Key Stage 1 (pages 68-69) provide opportunities to assess:
3.3a Be able to link the use of common materials to their simple properties.

 **HOUSEHOLD MATERIALS**

RESOURCES A collection of household materials including cooking oil, bleach, cleaning agents etc. Children should not handle harmful materials such as bleach

Class or group discussion The teacher in discussion with the children talks about what the materials are used for. Children could write 'rule books' about how and when they should be used and by whom.

The rule books should focus on where to get the information and how to use the materials, what precautions should be used and how first aid could be offered if accidents occur.

ASSESSMENT This activity develops understanding of the Programme of Study for this strand but provides no direct assessment opportunity.

4 LIQUIDS, SOLIDS AND GASES

RESOURCES A collection of solids, liquids, and pictures of objects containing gases

Research Children could find out about the differences between solids, liquids and gases.

Research could be carried out into what the properties of these things are and what makes them particularly useful for the jobs they do.

Practical Children should be given the opportunity to classify solids, liquids and gases.

ASSESSMENT This activity provides an opportunity to assess:
3.4a Be able to classify materials as solids, liquids and gases on the basis of simple properties which relate to their everyday uses.

5 THE SEWAGE FARM

RESOURCES A visit to a sewage farm

At the farm Children should find out how the water is cleaned for safe use.

Practical Once back in school the children could investigate ways of purifying dirty water, using different materials. The emphasis should be on the use of filtration as a method of purification.

Children can set up filtering systems by taping together two plastic bottles (one with the base cut out) and placing materials that they feel will filter dirty water, thus acting as a 'sieve' at the join (see Figure 2). Discuss how the dirty water will be 'produced' so that it is the same for everybody and comparisons can be made on the way different materials filtered the water.

Children could research water cleanliness in other countries and the need for careful use of water.

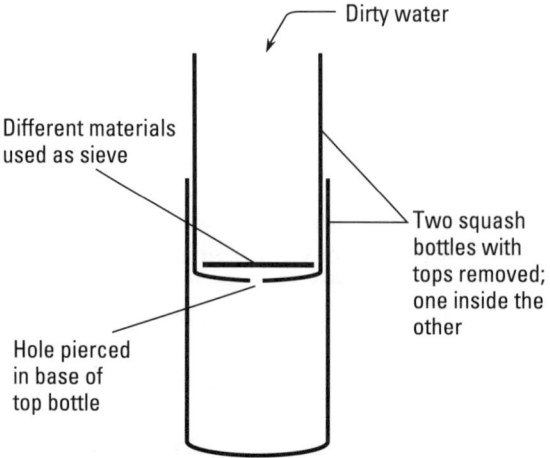

Figure 2 A simple filter apparatus

CHROMATOGRAPHY

RESOURCES A collection of black water-based felt-tip pens. Blotting or filter paper

Practical Write a number of messages on absorbent paper, preferably blotting or filter paper.

Take one of the messages and drip a small amount of water onto one of the words using a dropper. Show the children how the black spreads out and forms a range of colours. Explain that the range it creates is distinct to that pen.

Can the children match up the black felt-tip pen used to write the message?

This will involve the children creating the colour pattern for each pen and each message and then matching them up.

Discuss with the children the fact that the black ink in the pen is made up of a number of colours and that by dripping water onto it it breaks up into the mixture of colours that were used to create it in the first place. This splitting of colour into its component parts is called chromatography.

Children could research what they can about chromatography and its uses.

MIXTURES

RESOURCES Mixtures prepared by the teacher including mixtures of solids, liquids and liquids and solids. Useful mixtures include:

– sand and sugar;
– iron filings and sand;
– sugar and water;
– flour and water;
– olive oil and water;
– liquid soap and water.

Before this activity is undertaken children should have undertaken work to develop their understanding of filtration and evaporation (see next paragraph). The teacher should also have explained by demonstration how certain liquids can be separated by decanting. To demonstrate decanting it is important to have a container that pours without spilling and to have a steady hand.

Class demonstration

To demonstrate evaporation a saucer of water can be left at the side of the room and the time taken for evaporation to occur measured. The teacher can demonstrate how the rate of evaporation can be increased and used to separate mixtures by mixing salt into a small amount of water and then boiling until all the water has evaporated and the salt is left behind. It is important that the teacher undertakes the demonstration and not the children.

Class or group discussion Ask the children to explain what they know about the properties of each substance used in the mixtures in terms of where it comes from, whether it floats on or dissolves in water. Do not show the children the mixtures until after you have discussed the substances you have chosen to use. List all the properties they have identified for each substance.

Show the children the mixtures and discuss what they think each contains. Ask them to give reasons for why they think a certain mixture contains certain substances. Recap on the different methods the children have had experience of using to separate mixtures.

Practical

Once the children have discussed the mixtures ask if they can separate the mixtures. Methods of separating may include filtering, decanting and evaporating.

Children should use their knowledge of the properties of materials to help them.

Recording

Children should record the process they went through in order to separate the mixtures and explain how successful each method was. The children may well attempt to separate a number of mixtures so that they develop a greater understanding of the processes involved and how it is important that the most appropriate method should be applied to separating the mixtures. The most appropriate method can only be decided upon by drawing upon your knowledge of the properties of each substance in the mixture.

Class or group discussion

Children should be given the opportunity to discuss the most appropriate methods used for separating the substances. They should also be given the opportunity to relate the work to the use of the processes in industry and everyday life.

ASSESSMENT

Activities 5 to 7 provide opportunities to assess:
3.5a Know how to separate and purify the components of mixtures using physical processes.

ACIDS AND ALKALIS

RESOURCES

A collection of household acids and alkalis such as vinegar, lemon juice and bicarbonate of soda. A range of indicators which might include litmus paper, universal indicator paper or an indicator that has been produced by the teacher or children themselves

Making an indicator

Indicators can be made by gathering red cabbage leaves, elderberries or red petals from flowers. Mash up whichever you decide to use, then boil in water. Let the mixture evaporate slightly so that the colour becomes stronger and then use as a indicator. The indicator can be used in liquid form, for use on testing the acidity of soil, or soaked into blotting or filter paper which can then be used as an indicator for use in liquids.

Useful information

The sharp taste we get from biting into an apple or lemon is an **acid**. Acids always have a sour taste but we must not taste many of them as they are extremely harmful. Onions, oranges and limes contain acids.

Examples of common **alkalis** are washing powders, caustic soda, and ammonia solutions.

A **neutral** substance or solution is neither acid or alkaline as in the case of pure water and petrol.

Class or group discussion

Discuss with the children what an acid, an alkali and a neutral solution are and why it is useful to know. Explain where they are used in the home. Explain to the children that it is possible to make an indicator to tell you how acid or alkaline something is by using 'red cabbage water'.

Examine what happens to the liquid indicator when household solutions are added to it. The children should record what they observe.

'Red cabbage water' turns purple when an acid is added to it and green when an alkali is added.

Universal indicator paper turns red when in contact with acid and blue for an alkali. It will turn greenish yellow to indicate neutrality.

Litmus paper turns red in acid and blue in alkali.

ASSESSMENT	**3.5b** Be able to classify aqueous solutions as acidic, alkaline or neutral using indicators.

INVESTIGATION

"What affects the speed at which sugar granules dissolve in water?"

The properties, classification and structure of materials

Starting point
A beaker of sugared water which the children taste. A discussion about where the sugar granules have gone could follow.

Observing and asking questions
The focus of the discussion should centre around what has created the sweet taste in the water and why it is that the sugar can no longer be seen. Before the children carry out any investigations it would be very useful for them to explore what happens to a variety of substances when they are added to water. For example, flour, sugar, salt, vegetable oil, liquid soaps, food colourings, golden syrup and bicarbonate of soda. The whole emphasis of this initial activity would be on the children exploring and discovering that some substances dissolve when added to water whilst others do not. Words such as suspension could be introduced to explain those substances which do not dissolve. It is also interesting to see not only how solids react when added to water but also what happens to liquids, such as oil or washing-up liquid, that have been added to water.

A great deal of descriptive work and recording of observations could be done at this stage so that the children build up a source of information to consult when they begin to think of questions they would like to investigate at a later stage.

The children should be warned before any investigation begins of the dangers of tasting liquids unless they have been told that they are safe by an adult.

Resources
- granulated sugar
- brown sugar
- caster sugar
- icing sugar
- sugar cubes
- water
- kettle
- thermometer
- scales
- spoons
- heat-resistant containers
- beaker
- stop clocks

Children may ask
Does the colour of sugar affect how quickly it dissolves?
Does the amount of sugar affect how quickly it dissolves?
Does the amount of water affect how quickly the sugar dissolves?
Does the number of times the liquid is stirred affect how quickly the sugar dissolves?
Does the temperature of the water affect how quickly the sugar dissolves?
Does sugar dissolve more quickly than salt?

Predicting and hypothesising
The children should be able to make well-informed predictions about their investigations if they have undertaken a selection of Activities 1–8 from AT3(i) KS2. They may be able to make predictions which relate to the size of the granules of sugar. For example, *I think the icing sugar will dissolve more quickly than the granulated sugar because the granules are much smaller and*

the water will dissolve them quicker than the larger granules. They may even suggest that the hotter the water the faster the sugar will dissolve because they have noticed this fact when adding sugar to tea.

Designing and planning the investigation

The children should be thinking very carefully about how they are going to ensure their investigations are fair. They will also need to consider what will be appropriate masses of sugar and volumes of water to work with so that they obtain reasonable results. If the children select inappropriate quantities, the teacher will need to give careful guidance.

Type of sugar In this investigation the children need to investigate at least three different types of sugar. All other aspects of the investigation should be kept constant – the mass of each type of sugar, the volume and temperature of water, the number of stirs and the container in which the dissolving is to take place. The children will need to decide when to start timing. For example, will it be as soon as they start to add the sugar to the water or will it be when they start to stir? They will also need to decide what they mean by 'completely dissolved', so that they know when to cease timing.

Mass of sugar In this investigation the children will work with the same sugar type, temperature and volume of water, but they will change the mass of sugar. Careful choice of volumes should ensure that the results are significantly different to enable conclusions to be drawn and patterns to be identified.

Volume of liquid This investigation will require the children to keep everything constant other than the volume of water being tested.

Frequency of stirring This investigation will require the children to keep everything constant other than the number of times each sample is stirred.

Sugar and salt The properties of sugar and salt when added to water will be investigated. All other variables must be kept constant – the mass of sugar and salt, the volume and temperature of water, the number of stirs and the container.

Temperature of water This investigation will need to be carefully supervised if the children have decided to investigate some high temperatures. They should select at least three different temperatures and these should be significantly different to provide meaningful results. All other variables should be kept constant. If the children are having difficulty reading the thermometer then they can be shown. Inaccurate reading of the thermometer by the children may affect the patterns in the data they have collected.

Recording

As each of the investigations is likely to involve timing, the use of graphs, charts and tables as methods of recording and presentation will be appropriate. It is important that the children are encouraged to use the data they have collected to inform their conclusions later on. The children should be encouraged to present their graphs and charts in a way that can be understood by others and they should meet the conventions associated with the presentation of graphs and charts.

Drawing conclusions

Using the data that they have collected the children should be able to draw very accurate conclusions from these investigations. Wherever possible they should relate their conclusion back to their original prediction and ask themselves whether or not they were correct. They should also be encouraged to look closely at each investigation to see if it was fair. If it was not, could it have had an effect upon the results? Another important question they could ask themselves is, did they take enough readings? Would they have obtained more reliable data if they had taken three readings at each temperature and worked out the range and mean for each test?

The more able children may begin to wonder if there is more than one thing that affects the rate at which sugar dissolves in water. They may begin to pose such questions as *Does the temperature of the water have more of an effect on speed of dissolving than type of sugar?* In order to answer such questions the children will need to consider the data collected from both investigations and draw a conclusion based upon those results.

Attainment Target 3 Materials and their properties

Key Stage 1 strand (iii)
Chemical changes

PROGRAMME OF STUDY

A Materials, both manufactured and natural

Pupils should develop an awareness of which materials they are using are naturally occurring and which are manufactured.

B Exploring the effects of heating and cooling materials

They should explore the effects of heating some everyday substances, for example, ice, water, wax and chocolate, in order to understand how heating and cooling bring about melting and solidifying. They should observe materials such as dough, wood and clay which change permanently on heating.

STATEMENTS OF ATTAINMENT

3.2b Know that heating and cooling everyday materials can cause them to melt or solidify or change permanently (**B**).

3.3b Know that some materials occur naturally while many are made from raw materials (**A**).

ACTIVITIES Chemical changes

NATURAL AND MANUFACTURED MATERIALS

RESOURCES   A collection of natural and manufactured materials. The collection could include such objects as seeds, wood, rocks, plastic bottle tops and bricks

Class or group discussion   Look at the materials and discuss the effect of each object upon the children's senses. Children should describe the texture, colour, hardness, size and shape of each object. Encourage them to tell you what they know about each material in terms of what it is, where it has come from, its use and how it was made. Record their observations so that they begin to form a picture of each material.

Explain to the children that some materials are natural and others are manufactured. Explain what is meant by the terms and give examples that they will be familiar with.

Practical   The children are encouraged to sort the objects into groups according to whether or not they are natural or manufactured.

Children could find out about how the manufactured items have been processed. What is the natural raw material that the manufactured examples are made from?

Choose one natural material each week and ask the children to bring into school any manufactured products that have been made from it.

2 HOUSEHOLD OBJECTS

RESOURCES A selection of common objects found in the home

Practical Ask each child to bring into class one common object that is used in their home. Ask half the class to bring in an object made from natural materials and the other half to bring objects made from manufactured materials. Many parents will enjoy helping their children to find suitable objects.

Children discuss different objects found around the home. Are the objects made of natural materials that have been manufactured? Why are the objects made of the materials they are?

3 SNAP!

RESOURCES A pack of 'snap' cards, each with a picture of either a natural or manufactured object on the back

The teacher makes up two sets of cards – one set with pictures of manufactured objects on them and the other set with pictures of natural materials. Each card is placed on the table – the children shout *snap!* when the manufactured object matches the raw material from which it is made.

4 FABRICS

RESOURCES A selection of different fabrics, both natural and man-made

Class or group discussion Look at the fabrics and discuss the properties of each fabric using the children's senses. Children should be describing texture and colouring. Encourage the children to tell you what they know about each fabric in terms of what it is, where it has come from, what it is used for and how it was made. Record their observations so that they begin to form a picture of each fabric. The teacher will need to explain where certain materials come from and what they are made from.

The children could look in detail at each fabric sample using microscopes and hand lenses.

Children could describe the appearance of each fabric and then make a sketch of what they see.

The fabrics could form part of a wall display where the children are encouraged to group natural fabrics, and man-made fabrics.

ASSESSMENT Activities 1 to 4 combined provide opportunities to assess:
3.3b Know that some materials occur naturally while many are made from raw materials.

5 MAKING DOUGH

RESOURCES Flour, water and yeast for making into dough

Class or group discussion Discuss with the children each of the ingredients used to make dough.

Class demonstration Show the children how to mix their dough taking care to explain the health factors in keeping utensils clean and ensuring hands are clean before starting work.

Practical Children investigate what happens when equal portions of dough are baked for different lengths of time. The children discuss properties such as colour, texture and size. They could compare the inside of the dough to the outside.

JELLY CUBES

RESOURCES A collection of jelly cubes

Class or group discussion Give clear instructions how to dissolve the cubes of jelly. The children could try to explain what happens to the jelly and where it 'goes'. They could discuss how the jelly cube has changed.

The children could investigate how long it takes for jelly to set. They could place a bowl of jelly in different places (for example, cold, warm) to see if this affects the setting time.

CLAY

RESOURCES A supply of clay

Practical Children handle the clay and make simple shapes.

Leave the clay to set and then encourage the children to observe the changes that have taken place.

Has the size changed?

Some models could be fired so that the children could see the effect of heat on clay.

ASSESSMENT Activities 5 to 7 combined with the following investigation will provide opportunities to assess:
3.2b Know that heating and cooling everyday materials can cause them to melt or solidify, or change permanently.

INVESTIGATION Chemical changes

"How does heat affect wax, ice, chocolate and jelly?"

Starting point
A collection of solids which melt on heating.

Observing and asking questions
Discuss with the children what the solids are and encourage them to describe their properties. Drawings and sketches could be made to encourage the children to look carefully at the solids. The teacher and the children could discuss the simple properties of solids and compare them to liquids and gases. Discussion could also include what they think will happen to these solids if they are heated.

Children may ask
Do different solids melt at different speeds?
Does the amount of a solid affect how quickly it melts?
Does the amount of heat affect how quickly a solid melts?

Whenever children investigate using heat and how it affects the speed of melting, very close supervision will be needed. It should not be necessary to work with very high temperatures particularly if the children work with ice or chocolate. Warm water should be sufficient to begin the melting process.

Resources
- heat source
- kettle
- wax, ice, chocolate, jelly (and others of teachers' choice)
- suitable heat-resistant containers
- water
- safe area
- scales
- stop clocks
- thermometers

Predicting and hypothesising
Before carrying out the investigation encourage the children to predict what they think will happen. Most children will be able to relate their predictions to their everyday experiences of handling or observing melting ice, chocolate or jelly. Reasonably accurate explanations for their predictions should be provided by the children.

Designing and planning the investigation
When the children have decided what variable they wish to investigate encourage them to think carefully about how they are going to make their experiment fair. The children should be prompted to investigate one variable at a time.

Different solids Children could choose a variety of solids to investigate. Solids that melt could include: jelly, chocolate, ice and wax. Encourage the children to set up a fair test by ensuring that the mass of each solid is the same and that each solid is heated to the same temperature.

It is essential that this investigation is carefully supervised.

The children should decide what mass of each solid they are going to investigate and how they are going to heat the solid.

Different mass of a solid As with the previous test use solids that will melt readily. In this test it will be necessary to ensure that the solid being tested is the same and the temperature used is the same for each sample. The only thing that needs to be changed is the mass of solid. Careful teacher supervision is required. The children may need help in weighing their samples of solids.

Different temperature This test will be the most problematic to conduct as measuring different temperatures and keeping the test fair will be difficult. However if water is heated and the temperature measured before trying to melt the solid it can be done with careful teacher supervision. (The solids could be placed in plastic bowls and hot water could be placed in a larger bowl. The bowl with the solid in could then be placed on top of the hot water.) It is most likely that only the most able children in Key Stage 1 will undertake such a test. The solid used and the mass of solid must be kept constant; it is only the temperature that should be changed.

Children choose the equipment they need and then set up the investigation so that it can be conducted safely.

The children will need to make decisions about at what point they will call their sample 'melted'. The melting of each solid needs to be timed carefully to see if there are any significant differences.

Recording
Children need to decide how they will record their observations so that classmates will be able to understand their findings. Encourage them to record actual data rather than just simple observations.

Drawing conclusions
The children need to draw conclusions from their data and to give explanations. As mentioned earlier, it is highly likely in this investigation that the children's original predictions will have been well informed. Encourage them to look for patterns in the data they have collected. Encourage them to look for anything at all that they did not expect.

Attainment Target 3 Materials and their properties

Key Stage 2 strand (iii)
Chemical changes

PROGRAMME OF STUDY

A Where do materials come from?

Pupils should explore the origins of a range of materials in order to appreciate that some occur naturally while many are made from raw materials.

B Changing materials

They should investigate the action of heat on everyday materials resulting in permanent change. These might include cooking activities and firing clay. Pupils should explore chemical changes in a number of everyday materials such as those that occur when mixing plaster of Paris, mixing baking powder with vinegar, and when iron rusts.

C Energy and waste products

They should recognise that combustion of fuel releases energy and produces waste products including gases.

STATEMENTS OF ATTAINMENT

3.2b Know that heating and cooling everyday materials can cause them to melt or solidify or change permanently (**B**).

3.3b Know that some materials occur naturally while many are made from raw materials (**A**).

3.4b Know that materials from a variety of sources can be converted into new and useful products by chemical reactions (**B** and **C**).

3.4c Know that the combustion of fuel releases energy and produces waste gases (**B** and **C**).

3.5c Understand that rusting and burning involve a reaction with oxygen (**B** and **C**).

ACTIVITIES Chemical changes

CHANGES IN DOUGH

RESOURCES Flour, water and yeast for making dough

Class or group discussion Explain to the children the ingredients for making dough. Explain where each ingredient comes from and explain that by mixing the ingredients together each has an effect on the others. Explain these effects to the children.

Class demonstration Show the children how to mix their dough taking care to explain the amounts used and the health factors in keeping utensils clean and ensuring hands are clean before starting work.

Practical Children investigate what happens when equal portions of dough are baked for different lengths of time. The children discuss things like colour, texture and size. They could compare the inside of the dough to the outside.

Children should be developing an understanding that materials can be converted into new and useful products by chemical reaction.

2 CHANGES IN JELLY CUBES

RESOURCES A collection of jelly cubes, thermometers, mixing bowls, wooden spoons for stirring and hot water stored in a thermos flask

Class or group discussion Look at the jelly cubes and discuss with the children how they can dissolve them. Give clear instructions on how to dissolve the cubes. What volume of water will be used and how will they record the temperature of the water? How will they know when the cube has dissolved? What will they use to measure the time it takes for the jelly to dissolve?

This activity can be undertaken as a simple observational task which looks at how a solid dissolves and then resets; or it can be tackled as a simple investigation. If undertaken as an investigation it will introduce the children to the need for careful planning in setting up and carrying out an investigation. The children could explore whether the jelly melts at a different rate if the mixture is stirred. Each group could stir for a given length of time and results recorded and compared. One group will need to simply place the jelly in the water and not stir at all to act as a control.

Practical Use this investigation to encourage the children to consider how to set up a fair test:

– do we all have the same volume of water?
– is everybody's jelly the same size?
– how do we stir the mixture at the same rate?
– how do we decide when the jelly has dissolved?

The children melt the jelly using hot water. This requires close teacher supervision.

The children explain what happens to the jelly and where it goes. They also discuss how the jelly cube has changed.

The children could investigate how long it takes jelly to set. Is it the same wherever the jelly is put?

3 CHANGES IN CLAY

RESOURCES A plentiful supply of clay

Cross-curricular links
Art Children handle the clay and model simple shapes and objects. The children can describe how the clay feels. This simple investigation could be undertaken in an art lesson where children are introduced to the techniques for working with clay.

Encourage the children to note the differences between the clay before and after hardening. They should weigh the models as soon as they are finished and then draw simple sketches which include accurate measurements of the dimensions. The same measurements should be taken once the clay models have hardened. Have their models changed in any way? Encourage the children to record what they have observed and why they think the changes have taken place.

Children should be encouraged to link what they have discovered to examples around them. For example, cracks in plaster that appear on walls after a period of drying.

The children could go on to find out about the affect of heat on clay – by modelling and firing their own pots.

4 CHANGES IN BREAD

RESOURCES	A sliced loaf and a toaster with a timer
Class discussion	Discuss changes that occur when bread is toasted.
Practical	Toast a number of slices of bread for different periods of time.

Children are encouraged to observe closely, using a variety of senses, a slice of bread. What does it weigh? What is its area? How does it taste? The teacher could pose a question to encourage the children to discover what happens to the texture, taste, mass and area of a slice of bread after toasting. Toast slices of bread for different time periods and record the effect it has – through drawings and writing.

ASSESSMENT Activities 1 to 4 combined with activities at Key Stage 1 (pages 80-82) and the investigation for Key Stage 1 (pages 82-83) will provide opportunities to assess:
3.2b Know that heating and cooling everyday materials can cause them to melt or solidify or change permanently.
3.3b Know that some materials occur naturally while many are made from raw materials.

5 BURNING

RESOURCES	A burning candle
Class or group discussion	Discuss with the children what a candle wick needs in order to burn. The discussion could lead to children identifying the constituent parts of air.
Class demonstration	The teacher could demonstrate that oxygen is needed for the burning process to occur. This could be done by placing a jar over a burning candle. When the oxygen is used up the candle will go out.

The amount of oxygen in the jar can be estimated by placing a jar over the candle standing in a saucer of water. The height the water rises in the jar gives an indication of how much oxygen has been burnt.

ASSESSMENT This activity will provide opportunities to assess:
3.5c Understand that rusting and burning involve a reaction with oxygen.

6 CAR POLLUTION

Class discussion Discuss pollution caused by car exhaust gases.

Class demonstration The children research into how a car works and how the fuel that is used in cars is turned into energy to make the car move.

The research should also include how and what waste products are produced.

How are car manufacturers trying to reduce car pollution?

Also the children could write to car manufacturers to find out what is being done to reduce the effects of waste products from cars.

7 THE POWER STATION

RESOURCES A visit to a power station

The children could find out about what the power station does. What raw materials are used and in what form energy is produced. A great deal of research could be carried out on how and where the energy is used. Consideration needs to be given to the waste products produced by the power station.

ASSESSMENT Activities 6 and 7 provide opportunities to assess:
3.4c Know that combustion of fuel releases energy and produces waste gases.
Understanding of statement **3.4b** can be developed by carrying out the following investigation.

INVESTIGATION Chemical changes

"What affects the speed at which plaster of Paris sets?"

Starting point
A Plaster of Paris cast.

Observing and asking questions
Children could discuss the plaster of Paris cast and what it is used for. Where have the children seen them used? Why are they used? Discuss how they think the cast was made. Look at the powder and discuss with the children what they think is needed to make the cast.

Children may ask
Does the temperature of the water affect how quickly the plaster of Paris sets?
Does the amount of water used affect how quickly the plaster of Paris sets?
Does the quantity of plaster of Paris powder affect how quickly the plaster of Paris sets?

Predicting and hypothesising
Encourage the children to make a prediction about what they think will happen and why. Children at Key Stage 2 will have considerable knowledge to draw upon from work undertaken by Key Stage 1 and could be encouraged to relate their predictions to relevant prior knowledge. Where

Resources
- plaster of Paris mix
- water
- heat source
- suitable mixing bowls (not those used for food)
- old spoons
- plaster of Paris cast
- moulds
- stop clocks
- thermometers
- scales
- measuring cylinders

have they seen things before that solidify, and what seemed to affect the speed at which solidification took place?

Designing and planning the investigation
Encourage the children to set up a fair test. They should understand what to change and what to keep the same.

Water temperature Children need to decide appropriate masses of powder and volumes of water to investigate. These will need to be kept constant in each test. They will need to decide the range of temperature they will be investigating so that they get significantly different results. They will also need to decide how many times the mixture will be stirred.

Close supervision will be needed when water is heated.

Water volume Children need to decide appropriate masses of powder and the temperature of the water. These will need to be kept constant in each test. They will also need to decide what volumes of water they will be investigating so that they obtain significantly different results. They will need to decide how many times the mixture will be stirred.

Mass of powder Children need to decide appropriate masses of powder to investigate to ensure they obtain significantly different results. The temperature of the water will need to be kept constant in each test. They will also need to decide what volume of water they will be using in each test. They will need to decide how many times the mixture will be stirred.

The children will need to decide what equipment to use.

Decisions need to be made as to where the investigation should take place. Safety needs to be considered at all times – particularly with the use of hot water.

The children will need to decide what jobs individuals will undertake within the investigation.

Be careful not to pour any plaster of Paris mixtures down the sink.

Recording
At the planning stage of the investigation the children will need to decide at what point they consider the plaster of Paris to be 'solid.' This decision will have a bearing on the results that are recorded. Will the children simply give a constant time limit during which the mixture will have set or give criteria that the mixture must fulfil before it can be considered to be solid? For example, the point at which the plaster will not take a thumb print (this may be too demanding on time). The children will need to decide how well they can record their findings so that definite conclusions can be drawn.

Drawing conclusions
The children should relate the results to their original idea. Every effort should be made to encourage the children to relate their findings to the data they have collected.

From the data they have collected is it possible for them to predict what would have happened if they had made other measurements? For example, is it possible to predict the speed of setting for a mass of powder that has not been tested simply from reference to the patterns in the data the children have collected?

As follow-up work could the data the children have collected inform real-life situations in hospital? Contact your local hospital to find out how plaster of Paris casts are made.

Attainment Target 3 Materials and their properties

Key Stage 1 strand (iv)

The Earth and its atmosphere

PROGRAMME OF STUDY

A Observing rocks and soils

Pupils should observe and compare natural materials found in their locality, including rocks and soils.

B Weathering in the locality

They should observe the effects of weathering in their locality.

STATEMENTS OF ATTAINMENT

3.3c Understand some of the effects of weathering on buildings and rocks (**A** and **B**).

ACTIVITY The Earth and its atmosphere

WEATHERING

RESOURCES A walk around your locality

Cross-curricular link Geography This work could be incorporated into work on the locality in geography. As well as looking at weathering, opportunities should be taken to develop children's geographical vocabulary of features within the environment.

Practical Walk around an area near to school. No matter what type of locality you are sited in you will find effects of weathering. It is important that teachers have walked the route beforehand to identify signs of weathering. Look at the buildings closely, including the school, and at each construction material in turn. Old buildings provide particularly good examples of weathering, particularly churches.

Take this opportunity to identify geographical features that can be identified on the walk.

Walking around your locality what signs can the children see of weathering?

– timber rotting through damp;
– paint scaling off through heat from the Sun;
– cracks in walls due to movement underground or a building drying out;
– colour changes in paintwork due to sunlight;
– disfigured gargoyles or gravestones due to effects of wind and rain.

The teacher should help the children to identify the effects of weathering and identify reasons for it.

Children should be building up an understanding of:

- weathering taking place within their locality;
- different types of weathering around them;
- types of materials susceptible to weathering;
- how it is possible to prevent or slow down the effects of weathering;
- how weathering takes place – for example, that the wind can carry tiny particles that hit against materials and wear them down;
- that some construction materials are made from rocks which are weathered.

ASSESSMENT   **3.3c** Understand some of the effects of weathering on buildings and rocks.

INVESTIGATION   The Earth and its atmosphere

"How are rocks different?"

Starting point
Children could examine and handle each of the rock samples. Each rock should be clearly labelled with its name or a number to identify it.

Observing and asking questions
The teacher could explain as simply as possible how rocks are made. Explanation could be given that some are created in the sea, some form deep below ground to be released through volcanoes and others are changed by the effects of heat and pressure. The children should realise that all these rocks have names and some are used for building homes, larger buildings or walls.

They should be encouraged to look closely at each rock and begin to identify certain features. What are their first impressions of the rocks? What colour are they? They should ask and answer questions about how each rock sample feels in terms of its texture and mass. The children should describe each rock using their senses and could even give each a name according to the features they identify.

As they examine the rocks using their senses they should be encouraged to find ways of sorting and classifying the rocks. This could be done by colour, hardness, ability to soak up water, shape, texture.

Children may ask
Are all rocks the same hardness?
Do all rocks soak up water?
Do some rocks break more easily than others?

The teacher could talk to the children and explain that scientists use certain tests to identify different rocks.

The children should be encouraged to suggest their own tests to identify rock types.

Predicting and hypothesising
Because the children have had plenty of opportunity to handle and examine the rocks before planning their investigations they should be able to give good reasons for their predictions. For example, their reasons for saying that sandstone will be good at absorbing water may well come from their observations of its texture and what the 'grains' in the rock look like. Children may well have come across some rocks in their everyday life and will be able to recall information from these experiences, for example, rocks they have found on the beach in the holidays.

Resources
Samples of:
- slate
- limestone
- granite
- clay
- sandstone
- chalk
- pumice
- two-pence coins
- steel nails
- measuring cylinder
- water
- protective goggles
- strong plastic bags

Designing and planning the investigation

The teacher and the children could together make a list of the tests that they have decided to carry out. The teacher could describe briefly each one so that the children are clear what they have to do. Encourage the children to think whether the test will be fair or not. They should be prompted to consider how to ensure that they carry out each test in the fairest way.

Children should be working as a group, setting out their own tests and carrying them out as carefully as possible. They should be ensuring the tests are fair each time and that they are recording their results as they progress.

 **Hardness** This investigation needs to be very carefully supervised and children should wear protective goggles. The children take a sample of rock each and investigate its hardness by scratching. The children could use their fingernails; if this does not make a mark try a two-pence coin. The children need to ensure that it is the same person testing each sample and that they apply a constant pressure as near as is possible.

Absorbency Before carrying out this investigation the children will need to think carefully how they will judge how absorbent the rocks are. If all the water is to be absorbed they may have a very long wait – so decisions such as this need to be made before carrying out the test. Another factor involved is trying to ensure that each sample of rock to be tested is approximately the same mass so that the test is seen to be as fair as possible. The best way to carry out this investigation is to stand each sample of rock in the same volume of water for an equal time period and then to remove the rock. The volume of water left in the tray can then be measured. The children may decide to weigh the rock before immersing it in water and then re-weigh it after a selected time period. Both approaches are equally valid providing the children give the samples long enough to absorb some water.

 **Brittleness** This investigation needs to be carefully supervised and the children should be wearing protective goggles. The children can take a sample of rock in a strong plastic bag and drop the bag onto a hard surface, or have masses dropped onto it. (Do not use hammers to break the rocks.)

The children need to ensure that the rock samples or masses are dropped from the same height in order to make the test fair. The children could use part of a carpet roll middle to enclose the rocks to prevent pieces of shattered rock from being scattered. If none of the rocks break then the children should increase either the height from which the rocks are dropped or the mass which is being dropped on them.

Recording

The children should record their results clearly so that they can explain what they have done and can compare their results. They should be asked why they think their results have happened. Charts and tables would provide excellent ways of recording their findings.

Drawing conclusions

The children should be able to establish from their results what they now know about each sample of rock. They should be able to put the rocks into some basic order of hardness or absorbency. Like scientists the children should have begun to build up a picture of something about which they originally knew little.

The teacher should record the main findings of the class and make a display including the rocks.

Children should be encouraged to use the real names of the rocks wherever possible.

There may be buildings or walls constructed from the rocks you have been observing near to the school. If so, bring them to the attention of the children so that they begin to understand that rocks are used in construction. Perhaps their findings will help them to see why a particular rock was used for that construction, for example, that it is hard, that water cannot enter it. Likewise how the property of a rock may influence how it is used, for example, graphite for pencils, chalk for the blackboard and diamonds for drills.

Reference

An excellent publication for work on rocks is *Exploring Earth Science*, published by Northamptonshire Science Resources.

Attainment Target 3 Materials and their properties

Key Stage 2 strand (iv)
The Earth and its atmosphere

A Weather records and the water cycle

Pupils should have the opportunity to make regular, quantitative observations and keep records of weather and the seasons of the year. This should lead to a consideration of the water cycle.

B Rocks, minerals and soils

They should investigate natural materials (rocks, minerals, soils), sort them by simple criteria and relate them to their uses and origins. They should be aware of local distributions of some types of natural materials (sand, soils, rocks).

C Weather and soil

They should observe, through field work, how weather affects their surroundings, how sediment is produced and how soil develops.

D Major geological events

They should consider the major geological events which change the surface of the Earth and the evidence for these changes.

STATEMENTS OF ATTAINMENT

3.3c Understand some of the effects of weathering on buildings and rocks **(C)**.

3.4d Know how measurements of temperature, rainfall, windspeed and direction describe the weather **(A)**.

3.4e Know that weathering, erosion and transport lead to the formation of sediments and different types of soil **(C)**.

3.5d Understand the water cycle in terms of the physical processes involved **(A)**.

ACTIVITIES The Earth and its atmosphere

A WEATHER PROJECT

RESOURCES Weather recording equipment, including equipment to measure windspeed, wind direction, rainfall, temperature (minimum and maximum) and a method for describing the weather

A number of primary schools are now investing in a computerised weather station that will record changes in the weather over a period of time and print out the results in various formats. These should be purchased in addition to the more traditional methods of recording the weather to give the children as wide an experience as possible of recording changes in the weather.

Plan

Teachers should plan to undertake a weather project over a period of say one full term. This will form a major part of this strand.

**Cross-curricular links
Information Technology (application and effects)
Geography**

Comparisons can be made between the two methods of recording the weather. Many aspects of Information Technology are covered here (see Assessment below).

See assessment opportunities listed below.

Practical

All children should be given the opportunity to record the changes in the weather and record their findings.

By installing a computerised weather recording system children could gain understanding of both methods, and cover aspects of the Information Technology statutory orders.

This activity requires considerable forward planning by the teacher. An area should be set aside where children can choose equipment to help make their recordings. Children will need to be shown how to use all the equipment, especially any computerised system. The recordings are vital and it will help if they can be on display on a wall in the classroom so that children can see the changes in the weather over time. They will begin to discuss the recordings and will soon come to expect certain results and be aware of when recordings are incorrect or are unusual for that time of the year.

Recordings can be saved and shown to children the following year to see how the results compare.

ASSESSMENT

Information Technology
2b Use information technology for the storage and retrieval of information.
3c Collect information and enter it in a data base, and to select and retrieve information from the data base.
3d Describe their use of information technology and compare it with other methods.
4a Use information technology to retrieve, develop, organise and present work.
4f Review their experience of information technology and consider applications in everyday life.

Geography
Ge1 Geographical skills
2d Record weather observations made over a short period of time.
4d Measure and record weather using direct observation and simple equipment.
Ge3 Physical geography
2a Recognise seasonal weather patterns.

Science
3.4d Know how measurement of temperature, rainfall, windspeed and direction describe the weather.

THE WATER CYCLE

RESOURCES A diagram of the water cycle. Better still would be a video or television programme which explains the water cycle

The children should have studied weather changes over time in Activity 1 above. They should also have studied the processes of evaporation and condensation. They could have this explained as the classroom windows steam up one day.

Children should also have been introduced to the idea of cycles. (Food chains, page 64.) Once these have been explained children enjoy setting out cycles that they have thought of themselves.

Class or group discussion The teacher should explain that the water cycle is made up of four parts:

- Evaporation of water takes place from large surface areas of water and forms water vapour.
- Clouds are formed from this water vapour and are carried by air currents until they meet the cold temperatures over mountainous regions.
- Condensation occurs, water droplets form and water begins to fall, as rain.
- Water is then transported by rivers and streams back to the sea and the process begins once again.

Practical Children can draw their water cycle. More able children will be able to include specific mountains, seas and rivers based upon their own knowledge or work covered in geography.

ASSESSMENT **3.5d** Understand the water cycle in terms of the physical processes involved.

INVESTIGATION The Earth and its atmosphere

"What are the effects of wind on soils?"

Starting point
A discussion about pictures or videos of sand storms or soil being blown away on farmland such as the prairies of the USA.

Observing and asking questions
The initial conversation between teacher and children should focus around what they have seen on the video recording or in the photographs. The children should be asked to describe what is happening and give the reasons why they think it is occurring.

The teacher should talk to the children and explain how soils are made and their different origins. The children could carry out some research work of their own to find out more on these subjects. Through close observation and handling of a variety of soil samples the children should be able to describe the soil constituents and whether or not all the soil samples are the same. It is very interesting for children to examine soil samples in detail (wearing disposable gloves at all times) and list all the things they find in each sample. Careful attention should be paid to differences in colour and texture. All the time the children should be developing an understanding that soils are made up of particles from rocks and that different rocks create different types of soils. The teacher should explain that soil can be blown away and that sand

Resources

A variety of natural materials:

- silver sand
- sand
- soil (as many different types as possible)
- a section of turf
- clay
- gravel
- any other materials that can be purchased from a builder's yard
- grass seed
- balances
- seed trays
- hair dryer, fan or balloon pump
- magnifiers, microscopes and hand lenses.

dunes are spread by wind. Can the children think of any other ways in which soils and sands can be moved? (Mining and transportation are two examples). Make a list of their suggestions.

Children may ask

Will some soils be blown away more easily than others?
Will dry soil blow away more easily than wet soil?
Does plant growth affect how easily soil is blown away?

Predicting and hypothesising

Encourage the children before they carry out their investigations to make a prediction about what they think will happen and to give reasons why this will happen. The children may make predictions based upon their everyday experience. *The sand will blow away most easily because I have seen how it blows across the beach on a windy day.* Likewise, more able children may make a prediction which calls upon some prior knowledge about plants and soil. *The soil with least plants in will blow away most easily because it has fewer roots in to help bind the soil together.* Some children may refer to knowledge they have gained in Geography work where they have learnt that in some countries where the natural vegetation has been removed the soil has been eroded away by wind and rain. It is often useful for the teacher to record the children's predictions so that comparisons can be made with their findings.

Designing and planning the investigation

Encourage the children to suggest ways in which they will make their investigations fair.

Ensure the children are clear what they are setting out to investigate and the way they are going to record their results.

The teacher will decide what equipment will be available to the children and whether or not it will be set out so that the children can choose the most appropriate equipment and materials.

Different soil types In this investigation the children will need to keep everything constant except the type of soil. The mass of each soil sample, 'wind' direction and strength, length of time the 'wind' will blow for and the surface area of the soil open to the 'wind' need to be kept constant in all the tests. The children also need to decide how they are going to measure the extent of soil loss due to 'wind' – will it just be direct observation or measuring the mass of the soil before and after each test?

Moisture content In this investigation the type of soil being investigated, the mass of soil, speed and direction of 'wind' and surface area all need to be kept constant. The variable being investigated, moisture content, needs to be changed for each sample. In order to make this test fair, the soil type chosen needs to be dried thoroughly and then each sample needs to have a different volume of water added. The children should be encouraged to investigate at least three different volumes of water as well as a 'dry' sample to compare the effects.

Plant growth This investigation will take a much longer period to set up and investigate – but could provide very interesting results. The type of soil in each investigation needs to be kept constant – but each soil sample needs to have a different quantity of plant growth. This could be done by planting different amounts of grass seed in each sample – from very little in one, to a large amount in another. Once the children are happy that each sample has differing amounts of plant growth the soils could be allowed to dry off and then investigated using a hair dryer as a source of 'wind'.

Recording

The children may well come up with a variety of ways of recording their findings using the materials under study. If the children have collected data it would be quite possible to present the findings in the form of a graph. Photographs and video recordings could be used to record the effects of 'wind' on the different soils.

Drawing conclusions

Children should be encouraged to draw conclusions from their results and not what they thought or someone has told them.

Every opportunity should be made to link their conclusions to real life situations. How can the knowledge they have gained help towards thinking of methods used to prevent soils and sand from blowing away?

As follow up work the children could look at a map of Africa and identify the extent of the Sahara Desert. They could try to find out where it is spreading, in which direction and at what speed. What advice would they be able to offer to stop or slow down the growth of the Sahara? Their suggestions should be based upon their results. In many parts of the world this situation is occurring and a great deal of information could be researched on the subject.

Farmers in this country have tried to cope with such problems and it may be useful to write to an organisation such as the National Farmers Union who will be able to provide information on what has caused soil erosion in the past and how farmers are dealing with it now.

Attainment Target 4 Physical processes

Key Stage 1 strand (i)
Electricity and magnetism

PROGRAMME OF STUDY

A Electricity and its dangers

Pupils should be made aware of some uses of electricity in the classroom and in the home and the dangers of misuse.

B The effect of magnets on materials

They should explore the effect of magnets on a variety of magnetic and non-magnetic materials and consider their uses.

C Electrical activities

They should experience simple activities using bulbs, buzzers, batteries and wires and investigate materials to discover those which conduct electricity and those which do not.

STATEMENTS OF ATTAINMENT

4.1a Know that many household appliances use electricity and that misuse is dangerous (**A**).

4.2a Know that magnets attract some materials and not others and can repel each other (**B**).

4.3a Know that a complete circuit is needed for electrical devices to work (**C**).

ACTIVITIES Electricity and magnetism

 MAGNETS

RESOURCES A collection of magnets, both bar and horseshoe shaped. Ensure some of the bar magnets are coloured red and blue to show the North and South poles. A variety of different materials, some magnetic and some non-magnetic

It is important to make sure the children understand that magnets attract some materials but not others. Do not teach the children that magnets attract things that are made from metal.

Demonstration Show the group the magnets and explain that they are made of a special metal that attracts certain materials. Explain that the ends are called poles. Show that some magnets are stronger than others by seeing how many paper clips each can lift.

Show how the magnets are attracted to each other but that the same poles will not attract. Explain that they push each other away – they repel each other.

Practical Give the children a number of magnets and several materials in a container. Give them time to 'play' with the magnets and to feel the push and pull of the poles when brought together.

Ask the children to test which materials the magnets will attract. Ensure the children record their findings.

Gather the group together and see if everyone had the same results. Clarify, by demonstration, where children disagree so that it is clear which materials were and were not attracted by the magnets.

Set up a display where children can use magnets and reinforce their knowledge of the types of materials magnets do and do not attract.

ASSESSMENT **4.2a** Know that magnets attract some materials and not others and can repel each other.

---

2 ELECTRICITY IN THE HOME

RESOURCES A collection of common electrical items found in the home and at school

Class discussion Explain that each item needs electricity in order to work. Electricity is produced far away and enters our homes and schools through electrical wires and we connect our appliances to the electricity by plugging into a socket.

Ask the children to imagine how different our homes would be without electricity. Explain that electricity can be dangerous and so we need to take special care when using anything connected to electricity.

Practical Walk around school and spot the items that use electricity. Use the opportunity to explain how some of the items are used, for example, the photocopier. Explain the school rules for using anything electrical, for example, setting up the computer.

Class discussion Discuss ways in which they can ensure no accidents happen within their homes.

ASSESSMENT **4.1a** Know that many household appliances use electricity and that misuse is dangerous.

---

3 ELECTRICAL CIRCUITS

RESOURCES A collection of electrical items in shoe boxes

Organise the children to work in pairs and work with as many children as you feel confident with. Place the following items in the shoe boxes: short lengths of wire with the end stripped and exposed wire twisted, two 1.5 V batteries, two 2.5 V bulbs, a 6 V buzzer and a 1.5 V motor.

Explain that batteries are a source of energy. They store energy that can be used to make things work. They only have a certain amount of energy and cannot make every electrical appliance work but they have enough energy to work the items in the box.

Practical Describe the items in the box and ask the children to work in pairs to make the bulb light and the buzzer buzz and the motor turn.

Encourage children to help each other. Once they have discovered how to get an item to work then get them to explain to others how they did it. This will help to clarify whether they understand fully how to make a complete circuit or whether they managed to make an item work by accident.

Get the children to explain to others how they got each item to work. Ask the children to choose one item and to draw clearly the completed circuit.

Explain that when something works it is proof that they have made a complete circuit.

ASSESSMENT **4.3a** Know that a complete circuit is needed for electrical devices to work.

INVESTIGATION Electricity and magnetism

"How well do magnets attract?"

Resources
- a variety of different types of magnet
- paper clips
- rulers
- paper, card, plastic, wood, fabric, slate, Perspex

Starting point
A collection of different sized and shaped magnets.

Observing and asking questions
At the start of this investigation it is important that the children have had plenty of opportunity to handle and 'play' with a variety of magnets. Children should experience what the magnets can do and some of their simple properties. Discussion should initially centre around how the magnets are different. The children should be asked about how they recognise a magnet and what they know about the properties of magnets. The questions posed should encourage children to think carefully about what they have learnt from the activities already completed for this strand.

Children may ask
Which magnet is the strongest?
Will magnets attract through different materials?
Will magnets attract through different thicknesses of materials?

Predicting and hypothesising
Encourage the children to make a prediction about what they think will happen in their investigations and to give reasons why they think it will happen.

Predictions will be based very much on everyday experience. *I think the bigger the magnet the stronger it will be, because bigger things are always stronger.* More able children may well give reasons for their predictions which are based upon more advanced explanation – *The longer the magnet the stronger it will be because it has more 'magnetism' around it.* This prediction starts to show an awareness of the 'magnetic field' around a magnet and may well have come from some research the child has undertaken.

Designing and planning the investigation

The children will need to consider how they will carry out the investigation to ensure that they are as fair as possible.

Magnet strength The children need to decide what they consider makes a strong magnet. Is it the ability to attract many paper clips, or perhaps the distance away from the object to be attracted?

The children could use either of these ideas for measuring magnet strength. They could place each magnet they are testing into a container of paper clips and then count the number attracted to the magnet. The children will need to decide how long they are going to leave the magnet in the container of paper clips. Another approach could be to put one paper clip on a smooth table top; each magnet to be tested could be pushed towards the paper clip. When the paper clip moves, its distance from the magnet could be measured. The further the magnet is from the paper clip when it moves – the stronger the magnet.

Whatever the children decide upon it should be explained that they will all have to make sure they carry out the investigation in exactly the same way if it is to be a fair test.

Different materials In this investigation the children need to make sure that the magnet they are using is the same for each test but the materials they are testing are different. In order for this test to be fair, each of the materials being investigated should be of equal thickness – but all other variables should be kept constant. In order to observe definite differences between the materials children need to decide what constitutes attraction. *Will the magnet move three paper clips through every material?*

Material thickness In this investigation the children need to ensure that all variables are kept constant except the thickness of the material to be tested. The children should realise they are testing different thicknesses of the same material in all the tests.

Recording

Children will need to record their results in a way that can be easily interpreted by others.

Drawing conclusions

The children should be able to explain quite clearly which magnets were the strongest. The teacher should make sure that the children give responses based upon their results and not their original predictions.

Every effort should be made to encourage the children to compare their findings to their original predictions and for the children to ask themselves if some of their results were different from what they had expected.

Attainment Target 4 Physical processes

Key Stage 2 strand (i)
Electricity and magnetism

PROGRAMME OF STUDY

A Simple circuits

Pupils should have the opportunity to construct simple circuits.

B Using different components with electricity

They should investigate the effects of using different components, of varying the flow of electricity in a circuit and the heating and magnetic effects.

C Circuit diagrams and drawings

They should plan and record construction details of a circuit using drawings and diagrams.

D Electricity and safety

They should learn about the dangers associated with the use of mains electricity and appropriate safety measures.

E Magnets and materials

They should investigate the properties of magnetic and non-magnetic materials.

F Control and logic gates

They should begin to explore simple circuits for sensing, switching and control, including the use of logic gates.

STATEMENTS OF ATTAINMENT

4.2a Know that magnets attract some materials and not others and can repel each other (**E**).

4.3a Know that a complete circuit is needed for electrical devices to work (**A**).

4.4a Be able to construct circuits containing a number of components in which switches are used to control electrical effects (**B**).

4.5a Know how switches, relays, variable resistors, sensors and logic gates can be used to solve problems (**F**).

ACTIVITIES Electricity and magnetism

1 ELECTRICAL CIRCUITS

RESOURCES A collection of electrical items including 1.5 V batteries, battery holders, different lengths of wire stripped at the end and the wire twisted, 6 V buzzers, 2 V bulbs, bulb holders and 1.5 V motors

A collection of other useful items including lollipop sticks, small lengths of softwood that children can press drawing pins into easily, drawing pins, paper clips, Blu-Tack, Sellotape, scissors and easy-to-use wire strippers. Ensure you have explained how to use the wire strippers safely.

Remind the children that batteries are a source of energy and recap on work covered at Key Stage 1 (pages 98-101). Show the children the electrical items and explain what each item does. Demonstrate simple techniques for connecting wire to the different electrical items. For example, always twist the ends of the wire to make it easier to make a connection. Use small squares of thick card on which to mount the circuits. Each item can be stuck to the board using Blu-Tack. The children will find it easier to see the complete circuit.

Practical Ask the children to make a complete circuit incorporating one of the electrical items. Once they have done this ask them to make an accurate drawing of the circuit to enable others to make the circuit.

Class discussion Show a completed circuit to the class and ask them to think of a way that they could get the electrical device to go on and off. Most children will explain that you can just keep connecting and disconnecting the wire from the battery. Explain that this is a simple switch. Can they design their own more effective switch?

Children should be encouraged to think of different designs for switches. Introduce action words to describe how a switch works such as: push, pull, slide, press, turn, twist and drop switches.

Practical Give the children time to add a switch to their circuits. It is useful to limit the items they can use to those listed in the Resources section. It helps with organisation and ensures the children are not overwhelmed by having too many items to choose from.

Children should be given time to demonstrate their switches and for others to suggest ways in which they might be improved.

Now ask the children if they can add more than one electrical item to their circuit.

Group or class demonstration Children may ask to use more than one battery in their circuits as they consider this will give them more energy to make additional items work. This should be encouraged but first teachers should show the effect of attaching more batteries to a circuit which is to light a bulb. As batteries are added, the bulb shines more brightly but eventually it goes out as the heat created by the electricity was too great and has 'burnt out' the thin wire in the bulb. Once the thin wire (that the children can easily see in the bulb) has been broken, the circuit is no longer complete and does not work. Count up the number of batteries required to 'blow' a bulb and point out that the children should use fewer batteries to prevent the bulbs from being blown.

Safety Explain to the children that working with electrical items connected to 1.5 V batteries is not the same as working with mains electricity which can be very dangerous. Find time to discuss with the whole class the dangers of mains electricity.

ASSESSMENT **4.3a** Know that a complete circuit is needed for electrical devices to work.
4.4a Be able to construct circuits containing a number of components in which switches are used to control electrical effects.

2 'MAGNETS ATTRACT'

RESOURCES A range of magnets of different strengths and sizes. A range of materials that the children will be able to identify. Ensure there is a good mix of magnetic and non-magnetic materials

Practical Set up a display in the classroom where children can be given the opportunity to classify the materials into two groups: 'magnetic' and 'non-magnetic'. There should be an opportunity for the children to record their findings. Begin with simple materials and after a while add new materials, some of which should be unfamiliar to the children.

Place several different types of magnets on the display and ask the children to investigate whether magnets can attract other magnets.

Class discussion Once all the children have had the opportunity to work at the display, time should be found to talk with the class about their findings.

ASSESSMENT **4.2a** Know that magnets attract some materials and not others and can repel each other.

3 RELAYS AND LOGIC GATES

RESOURCES 2.5 V bulbs, 1.5 V batteries, battery holders, strips of insulated wire with ends bared, variable resistor and light-dependent resistor (LDR)

Variable resistor This is used to control the current in a circuit. It operates by introducing a change of resistance into a circuit between low and high, thereby altering the current flowing (see Figure 3). Variable resistors are used in dimmer switches to control the brightness of the bulb.

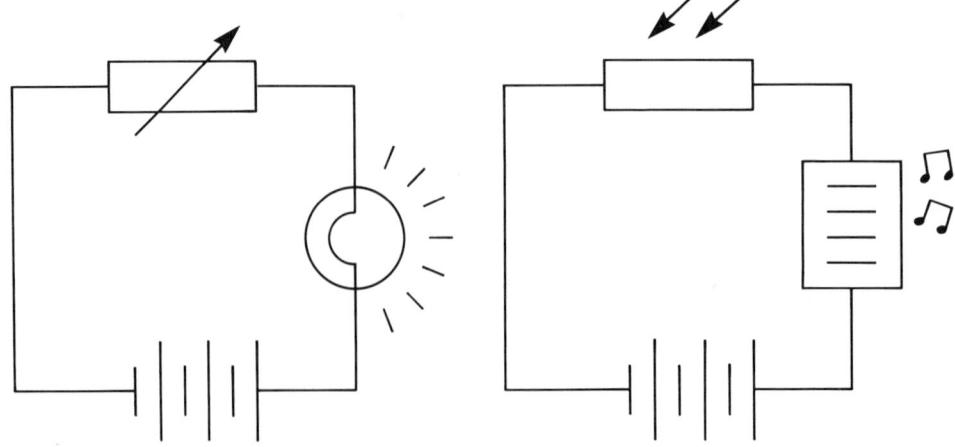

Figure 3 A variable resistor and a bulb in a circuit (left), and a LDR and a buzzer in a circuit (right)

Light-dependent resistor (LDR)

This changes its resistance as the level of light changes. When it becomes darker the resistance of the LDR increases. This reduces the current flowing in the circuit and the bulb dims or goes out. They are used to activate street lights or lighthouses when it gets dark. They are more commonly known as light sensors.

Logic gates

There are three types of logic gates: AND gates, OR gates and NOT gates. An AND gate is two (or more) switches in one circuit. When both are connected the bulb will light up (see Figure 4). An OR gate is where a bulb is connected to more than one circuit and the completion of any one circuit will light it.

Figure 4 AND gate, NOT gate and OR gate in circuits

Further information

A NOT gate is more difficult to explain and set up in a primary classroom as it can require items not normally found in primary schools. A NOT gate is a circuit which is complete but is waiting for the circuit to be broken before it lights the bulb (see Figure 4). They are found in a home where a burglar alarm has been installed. At night before going to bed, the alarm system will be activated. All the circuits are complete. As soon as a burglar forces a window the circuit is broken and the alarm is sounded. The breaking of the circuit causes another circuit to activate and sound the alarm. This is the best way of explaining a NOT gate to young children.

If you wish the children to devise their own NOT gates then the easiest way is to introduce the concept through computer control.

Note
Work involving LDRs and AND, OR and NOT gates is best undertaken through computer control work otherwise complex equipment not normally available in primary schools will be required.

Practical

Ask the children to explore the effect of adding a variable resistor to a circuit containing a bulb. Discuss ways in which they could be used in the home, school or in our daily life.

ASSESSMENT

4.5a Know how switches, relays, variable resistors, sensors and logic gates can be used to solve problems.

INVESTIGATION   Electricity and magnetism

It is suggested that it is inappropriate to undertake investigations within this particular strand and that children should (as set down in statutory orders) be given the opportunity to solve problems.

Teachers might like to combine this work with work being undertaken in design and technology or computer control work.

Suggested problems

1. Devise a lighthouse light that lights in the dark and goes off during daylight. (Uses an LDR.)
2. Use computer control to design and make a burglar alarm system that will sound if a burglar breaks the circuit. (Uses a NOT gate.)
3. Design a circuit that activates an illuminated sign only when two switches are pressed. (Uses an AND gate.)
4. Design and make a circuit that will allow us to control the amount of light in a baby's bedroom. (Uses a variable resistor.)
5. Design a burglar alarm system that sets off an alarm when the window is opened and the video recorder is removed. (Uses an AND gate.)
6. Design a burglar alarm system that sounds an alarm when the window is opened or the video recorder is removed. (Uses an OR gate.)

The burglar alarm scenario can be adapted to give children the opportunity to incorporate any of the devices listed within the Statement of Attainment **4.5a**.

7. Using a computer control system, write a procedure for lifting and lowering a railway barrier. The children can construct the barrier (using Lego card 19 from set 1033 and the card in Lego Technic 2 kit) which can then be connected to their computer control system (see computer control for further details).

Background information

Computer control

There are a number of control systems available but the list of equipment in Appendix 1 has been found to be particularly suitable for use in primary schools. An old BBC computer system is recommended that could be dedicated to computer control so that equipment can be left connected and children can have opportunities for writing simple procedures to control lights and switches during any spare time they may have. Computer control work will cover a great deal of the Programmes of Study in Information Technology, especially those under Applications and Effects and Measurement and Control (see assessment opportunities below).

ASSESSMENT

Science
4.4a Be able to construct circuits containing a number of components in which switches are used to control electrical effects.
4.5a Know how switches, relays, variable resistors, sensors and logic gates can be used to solve problems.

Information Technology
Measurement and control
3b Give a sequence of direct instructions to control movement.
5b Understand that a computer can control devices by a series of commands, and appreciate the need for precision in framing commands.
Applications and effects
3d Describe their use of information technology and compare it with other methods.
4f Review their experience of information technology and consider applications in everyday life.

Attainment Target 4 Physical processes

Key Stage 1 strand (ii)
Energy resources and energy transfer

PROGRAMME OF STUDY

A Fuels at home and school

Pupils should find out about the fuels used in their home and school.

B Hot and cold

They should talk about when and why they feel hot or cold and link the sensations of hot and cold with thermometer readings, for example in water and air.

STATEMENTS OF ATTAINMENT

4.2b Understand the meaning of hot and cold relative to the temperature of their own bodies (**B**).

4.3b Know that there is a range of fuels used in the home (**A**).

ACTIVITIES Energy resources and energy transfer

ICE CUBES

RESOURCES Ice cubes and Thermos flask for ice storage, thermometers and plastic containers

Class discussion A class discussion about when and why we sometimes feel hot or cold could be an opportunity to show the children a simple thermometer. (Programmes of Study indicate that the children need to link feelings of hot and cold with thermometer readings.)

Demonstrate how the coloured liquid moves on the scale when in contact with hot and cold water and air. The air temperature could be marked with tape or pen and compared between the outside and in the classroom. The children might suggest ways to make the readings move higher and lower on the scale. For example, holding the thermometer in the hands, blowing on the bulb or placing it in the refrigerator.

The teacher may feel it is appropriate to show how a numerical reading can be made and to introduce the terminology 'degree' as a unit of temperature.

What changes happen outside on very cold days?

Practical Containers of ice cubes are needed for the children to handle.

How does the ice feel? What happens to the ice as you hold it in your fingers? What is the ice cube made from? How do you know?

What will the ice cube do to a glass of orange or water?

The children could time how long it takes for an ice cube to melt. The temperature of the freezer compartment in the refrigerator could be recorded.

How cold must it be outside before ice forms?

2 HOT AND COLD WATER

Class discussion   What do we mean by 'hot' and 'cold'?

The aim of this investigation is to help children to understand that the meaning of hot and cold is relative to the temperature of their own bodies.

Organise three bowls of water:

1. cold water;
2. tepid water;
3. warm water.

Start with your left hand in the cold water and your right hand in the warm water. After about one minute place both hands together into the tepid water. How does it feel?

Practical   The children could be asked to make commonly used words such as hot, warm, tepid, cool and cold more 'scientific' by deciding on a definite temperature value, or range of values, for each term.

ASSESSMENT   Activities 1 and 2 combined provide an opportunity to assess:
4.2b Understand the meaning of hot and cold relative to the temperature of their own bodies.

3 FUELS

Class discussion   The children could ask the caretaker how the heating system works and what has to be done to make sure it operates. How is our school heated?

The network of pipes or vents could be discussed.

Where does the heat come from? What makes the heat?

The children need to be taught that there are a variety of fuels used to heat our homes and buildings. A class discussion should provide information about popular fuels used in homes and in school.

Reference material, especially videos, could introduce the children to where these different fuels come from.

What must happen to all these fuels in order to ensure warm air at home and in school?

Can any of the fuels be used more than once?

ASSESSMENT   Activity 3 provides an opportunity to assess:
4.3b Know that there is a range of fuels used in the home.

INVESTIGATION

Energy resources and energy transfer

"What affects how hot or cold we are?"

Resources
- thermometer strips
- thermometers
- a collection of warm winter clothes
- a collection of summer clothes
- stop watches

Starting point
A discussion about the clothes we wear and why certain clothes are worn at certain times of the year.

Observing and asking questions
The discussion about which clothes are worn at what times should focus on differences in temperature between winter and summer and why certain types of clothes are worn at particular times of the year. The children could be asked about ways in which they keep cool and ways in which they warm themselves when they are feeling cold. They could talk about those parts of the body that seem to get hottest or coldest and what they wear to account for these differences. There could be more discussion about how the body tells us when it is too hot or cold by the use of sweating and shivering. The children could be involved in some research work to find out what is happening when this happens. The children may enjoy going out on a cold day, just for a few moments, without a coat on to see which parts of their bodies feel cold soonest. Pictures of athletes who are very hot after exercise could be compared with pictures of Antarctic explorers or mountaineers to see differences in clothing.

Children may ask
Are all parts of our body the same temperature?
Which parts of our bodies cool down quickest?
How quickly do our hands warm up?
What type of gloves warm our hands best?
When we exercise do we get warmer?

Predicting and hypothesising
Before the children begin their investigations encourage them to make predictions about what they think will happen and why. If they have been involved in the class discussion about the clothes and the parts of the body which feel cold or feel warm they should be able to give well informed predictions. The reasons for the predictions will generally be based on everyday experience, for example: *I think that the gloves will warm up my hands when they are cold because I have to wear gloves in the winter time.* Another prediction might be: *I think our hands will be the coolest part of our body because they are furthest away from our main body.*

Designing and planning the investigation
The children should be encouraged to make the investigations fair. They may need to be shown how to use thermometers and thermometer strips in order to obtain accurate results.

Parts of the body In this investigation the children could measure the temperature of various parts of the body. They will need to decide which parts to measure. They should be encouraged to draw up a list of the parts and to devise a record sheet on which to write their findings. The parts should be easily accessible so as to give more accurate readings, for example, forehead, back of hand, leg and under the arm.

The children should take readings from a number of other children in order to get a fair selection of results.

Parts of the body that cool down quickest In order to carry out this investigation the children will need to decide which parts of the body they are going to study. They will need to record the temperature of each part of the body under normal conditions inside the classroom. Once this has been done the child or children being investigated will need to expose themselves to a cold environment. This will have to be handled carefully so as not to cause the child concerned any distress. The length of time the child is exposed to the cold needs to be carefully monitored. The children quickly record the temperatures of each part at the end of the time period. The temperatures before and after exposure to the cold should then be compared.

Hands In this investigation the children must take the temperature of their hand before cooling so they know the 'normal' temperature of their hand. The child should then hold an ice cube for a few minutes to cool their hand down. This must be carried out carefully so as not to cause any distress to the child. Once the hand has cooled, the other children record the rise in temperature of the hand. From this it should be possible to calculate how long it takes the hand to return to 'normal' temperature. Several children should be investigated to see if there are any significant differences between individuals.

Gloves The question to ask here is do certain types of gloves warm up hands quicker than other types? In order to find out the answer to this question it will be necessary to record the 'normal' temperature of the hands of several children. These children should then cool their hands on ice cubes for a defined period of time. Each child should then put on a different glove, for example, woollen, plastic, or cotton, and then the temperature of each child's hand should be recorded until the temperature has returned to normal. It should be possible to see which glove helps the hand to return to normal temperature quickest. It would be interesting to ask children at the end of this investigation if they think anything else might have affected the result.

Exercise The children in this investigation need to take the temperature of the individuals to be observed before exercise. These data need to be recorded and then a defined time period of exercise should take place. At the end of the exercise, the temperature should be taken again to see if there has been any significant alteration. The interesting aspect of this investigation will be to note differences in temperature between each child. An extension of this investigation might include seeing if lengthening the time period of exercise increases the temperature in proportion.

Recording
There will be a good deal of data resulting from these investigations, which can be recorded in a variety of forms. The children should try to refer closely to their data when they are drawing conclusions about their investigations.

Drawing conclusions
Encourage the children to look for any pattern in the data they have collected that will support their conclusions. The children should be encouraged to look closely at their results and compare their results to their original predictions.

Attainment Target 4 Physical processes

Key Stage 2 strand (ii)
Energy resources and energy transfer

PROGRAMME OF STUDY

A Movement

Pupils should investigate movement using a variety of devices, for example, toys and models which are self propelled or driven and use motors, belts, levers and gears.

B Heating and cooling substances

They should investigate changes that occur when familiar substances are heated and cooled, and the concepts of 'hot' and 'cold' in relation to their body temperature.

C Fuels

They should survey, including the use of secondary sources, the range of fuels used in the home and at school, their efficient use and their origins. They should be introduced to the idea that energy sources may be renewable or non renewable and consider the implications of limited global energy resources.

D Energy transfer

They should be introduced to the idea of energy transfer.

STATEMENTS OF ATTAINMENT

4.2b Understand the meaning of hot and cold relative to their own bodies (**B**).

4.3b Know that there is a range of fuels used in the home (**C**).

4.4b Understand that an energy transfer is needed to make things work (**A** and **B**).

4.5b Understand that energy is transferred in any process and recognise energy transfers in a range of devices (**A** and **D**).

4.5c Understand the difference between renewable and non renewable energy resources and the need for fuel economy (**C**).

ACTIVITIES Energy resources and transfer

 LEVERS

Useful background information can be found in the booklet *Forces – A Guide for Teachers,* NCC, 1992 (see page 118).

A lever is a simple machine which uses a pivot. A force is applied to one end of a lever. The greater the length of the lever the less force is required.

RESOURCES

The teacher could bring a collection of everyday examples of levers into the classroom, or which have a lever as an essential component to prompt discussion. For example, drink cans, tin of shoe polish, tin opener, poker, knife and fork. The collection should include levers of different lengths

Class discussion

What does a lever do?
Where are there levers in the classroom? (Examples are door and window handles.)
How do levers make jobs easier? How do they work?

Practical

Select the relevant cards for Lego Technic 1 and 2 introducing the principles of levers to the children. Allow the children to work through the selected cards to build up their knowledge of levers and their use in everyday situations. Ask the children to make their own levers using everyday materials, wood or Lego.

How do levers help people in different jobs?
Make a collection of pictures from magazines which show levers.

2 GEARS

Practical

Use the Lego Technic cards (see Appendix 1) that focus upon gears, or use *First Gear* to give the children the opportunity to investigate gears and their uses.

Examine how one gear affects the movement of another.
How can direction and speed be changed?
Devise a gear system which will move a roundabout.
How could gears be used to lift or move a heavy load?

The Lego Technic Cards introducing belt drives (set 1031, card no. 5) are another aspect of energy transfer which can be investigated.

Have belt systems got any advantages or disadvantages compared with gears?

ASSESSMENT

Activities 1 and 2 are starting points for investigations which may provide scope for assessment to be made of:
4.4b Understand that an energy transfer is needed to make things work.
4.5b Understand that energy is transferred in any process and recognise energy transfers in a range of devices.

3 HEAT

Class discussion

How do we keep the heat within our homes?
Which is the warmest part of your house?
Which parts of your house are the coldest?
How is the heat shared around the house?
How is heat sometimes lost?

Class or group discussion

The different ways of retaining heat within a building could be discussed and the teacher could show the children some of the popular insulation materials and devices now used in many homes. For example, loft insulation material, plastic lagging for pipes, polystyrene for cavity walls, draught excluders for windows and doors, double glazing.

Practical Why is the insulation of homes and schools very important? (Rising cost of fuel and limited global energy sources could be discussed.)

The children could investigate the effectiveness of wool, polystyrene, oil, foam and cotton wool by seeing which materials keep a hot drink from cooling quickly.

Suggestions for saving the school's heating bills might be offered.

ASSESSMENT **4.3b** Know that there is a range of fuels used in the home.
4.5c Understand the difference between renewable and non renewable energy resources and the need for fuel economy.

INVESTIGATION

Energy resources and energy transfer

"What affects the distance an elastic band cotton-reel motor can travel?"

Starting point
The teacher shows the children how to make an elastic band cotton-reel motor (see Figure 5).

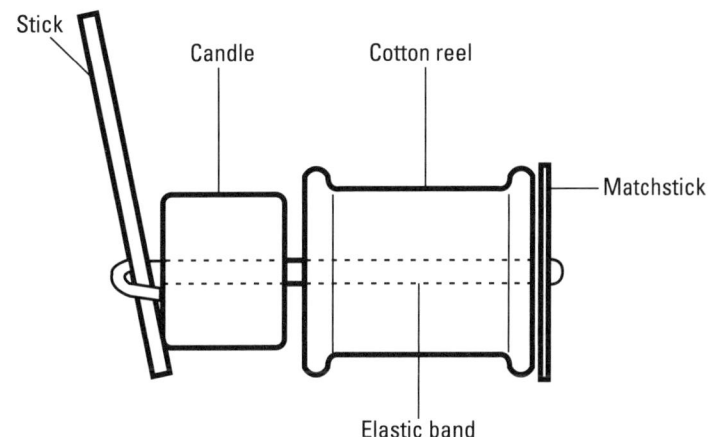

Figure 5 A cotton-reel motor

Resources
- cotton reels
- elastic bands
- match sticks
- candle wax
- rulers
- stop watch
- collection of toys which use stored energy to move

Observing and asking questions
The teacher will need to explain to the children the concept of energy being the reason for the motor's power and ability to produce movement. A motor contains stored energy and helping the children move towards understanding the transfer of that energy into movement is the main purpose of this investigation. Children and teacher could discuss where the energy is stored and how it is released. More discussion and questions could arise through examination of the collection of toys which use stored energy. The main focus of discussion could be on where the energy is stored and how the energy is released.

Children may ask
How does the number of 'winds' of the cotton-reel motor affect how far it will travel?
Does the size of the cotton-reel make any difference to how far it travels?
Does the length of the pusher affect how far the cotton-reel travels?

Predicting and hypothesising
Before the children carry out their investigations they should make predictions about what they think will happen and why. Their predictions are

likely to be related to their experiences with clockwork or electrical toys where they will be aware that the more 'winds' a toy has the longer it will run, or in the case of electrical energy – the larger the battery the longer it will last, or the more power it will have.

Designing and planning the investigation

'Winds' number In this investigation the children simply have to make sure that the same cotton-reel motor is used for each test and that the test is carried out on the same surface each time. The variable they are studying is the number of 'winds' – so these should be changed in a straightforward manner. Everything else should be kept constant.

Size of cotton-reel As it is often difficult to obtain different sizes of cotton-reel, the best way to approach this investigation is to make a variety of different sized cardboard 'wheels' that can be attached to the sides of the cotton-reel. By doing this it will increase the size of the cotton-reel by a standard amount each time. All other variables should be kept constant, for example the surface and the number of 'winds'.

Length of pusher In this investigation everything is kept the same apart from the pusher which is changed for each test.

Recording

Simple graphs could be used to record the number of 'winds' and the distance travelled. The children could write guides for younger children giving instructions on how to make the models.

Drawing conclusions

The most interesting part of this investigation will be in the close examination of the results obtained as it may well be possible to find interesting patterns in the data collected. The children should be encouraged to look for these patterns. By careful examination the children can advise other cotton-reel motor builders on the best size of reel and best length of pusher.

Attainment Target 4 Physical processes

Key Stage 1 strand (iii)
Forces and their effects

PROGRAMME OF STUDY

A Moving things

Pupils should have early experience of devices which move.

B Forces

They should experience the natural force of gravity pulling things down and manufactured forces such as those produced in wind-up toys, elastic or electrically driven toys and by the movement of their bodies. These forces should be experienced in the way they push, pull, make things move, stop things and change the shape of objects. Such experiences could include, for example, road safety activities.

C Floating and sinking experiences

They should explore floating and sinking and relate their experiences to water safety.

STATEMENTS OF ATTAINMENT

4.1b Understand that things can be moved by pushing or pulling them (**B**).

4.2c Understand that pushes and pulls can make things start moving, speed up, slow down or stop (**B**).

4.3c Understand that forces can affect the position, movement and shape of an object (**B** and **C**).

ACTIVITY Forces and their effects

PUSHING AND PULLING

In order to develop an understanding of forces young children need to be given a wide range of activities throughout Key Stage 1.

RESOURCES   Water tray containing a variety of objects. Collection of toys

Practical   Children should have opportunities to work at the water trays. They should develop an understanding that some objects sink, some float and others are suspended in water. Children should understand that mass and shape can affect whether an object floats or sinks.

They could explore how objects that float can be made to sink and those that sink can be made to float.

Make a collection of toys that can be moved in different ways. Divide the toys into two groups – those that move by pushing and those that move by pulling. Look at the ways other toys are made to move. Can we think of ways of classifying these other toys? (Winding, dropping, using springs, by elastic band.) Children could use their knowledge to help them to design and then make a toy in design and technology.

Cross-curricular links
Physical Education

Offers many opportunities for children to be involved in pushing and pulling activities. Specific lessons could be planned to develop a greater understanding that pushes and pulls can make things move, speed up, slow down, or stop. They should understand that forces affect the position, movement and shape of an object, and that a force is required to move an object and the greater the force the greater the movement.

Dance and work with small apparatus offer most opportunities to develop an understanding of forces.

Develop an understanding that different objects that float do so at different levels. Children should be given the opportunity to relate their experiences to water safety.

ASSESSMENT

4.1b Understand that things can be moved by pushing or pulling them.
4.2c Understand that pushes and pulls can make things start moving, speed up, slow down or stop.
4.3c Understand that forces can affect the position, movement and shape of an object.

INVESTIGATION   Forces and their effects

"How does a parachute slow down the speed at which a parachutist falls to the ground?"

Resources
- Plasticine
- fabrics
- sheets of paper
- tissue paper
- scissors
- measuring tapes and rulers
- stop clocks
- balls
- feathers
- small stones

Starting point
Exploration and discussion about how certain things fall to the ground. This could be accompanied by a selection of photographs or a video of a free-fall parachutist at different stages during descent.

Observing and asking questions
After playing with a variety of objects and watching how they fall and discussing the differences in how a free-fall parachutist descends, the children could be asked to consider how they feel they could slow down the speed of descent of a ball of Plasticine. The teacher could draw attention to the use of the word 'gravity' and encourage the children to give explanations of what they think it is. The teacher could show the children the difference between how a screwed up ball of A4 paper falls to the ground as opposed to an open sheet.

Encourage the children to try and explain why the latter falls more slowly.

Children may ask
Will a large parachute be better than a small one?
Will the material the parachute is made from affect the speed of descent?
Does the height from which the parachute is dropped make any difference to the speed of descent?
Will a larger ball of Plasticine fall more quickly than a smaller one?

Predicting and hypothesising
Nearly all children will be aware that a parachute is used to slow down the

speed at which a parachutist falls to the ground – what they will not know is how the size, or the material of the parachute affects the speed of descent. Encourage the children to make a reasoned prediction before beginning their investigation. These may well take the form of: *I think the larger the parachute the slower the ball of Plasticine will fall because it will catch more air to slow it down.*

Designing and planning the investigation

Parachute size In this investigation the size of the ball of Plasticine must remain constant as must the height from which it is dropped. The material that the parachute is made from must also remain the same – but the area of the parachute must be changed for each test. A problem the children may encounter is that the time taken for the parachute to fall may be so short that it is difficult to time with the stop clocks available in school. If this is the case the children will have to think carefully about the mass of Plasticine they are using and the height from which it is dropped.

Type of material In this investigation the size of the parachute and mass and shape of Plasticine used remains the same – it is just the material that the parachute is made of that changes. The drop height should remain constant.

Drop height Great care needs to be taken with this investigation and it is better if an adult supervises. The children will want to drop their parachute from a variety of heights and it is safer if an adult is involved in perhaps climbing a step ladder to get the extra height. All other aspects of the investigation need to be kept constant.

Mass of Plasticine In this investigation the mass of the Plasticine should be varied and all other variables should be kept constant.

Recording

A great deal of data will be collected from this investigation and can be presented in graphical form.

Drawing conclusions

The children should look carefully at the data to see if their original predictions were correct and wherever possible they should be encouraged to look for patterns which support their findings.

Attainment Target 4 Physical processes

Key Stage 2 strand (iii)
Forces and their effects

PROGRAMME OF STUDY

A Measuring forces

Pupils should explore different types of forces including gravity and use measurements to compare their effects in, for example, moving things and bridge building. They should investigate the strength of a simple structure.

B Forces working together

They should be introduced to the idea that forces act in opposition to each other, that one force may be bigger than another, or equal to it, and that the relative sizes and directions of the forces can affect the movement of an object. They should investigate the factors involved in floating and sinking. They should explore friction and investigate the ways in which the speed of a moving object can be changed by the application of forces. This work should be set in everyday situations, for example, road safety, transport (including cycling and sailing), balancing systems and hydraulic mechanisms in model making.

STATEMENTS OF ATTAINMENT

4.2c Understand that pushes and pulls can make things start moving, speed up, slow down or stop (**A** and **B**).

4.3c Understand that forces can affect the position, movement and shape of an object (**A** and **B**).

4.4c Know that more than one force can act on an object and that forces can act in different directions (**A** and **B**).

4.5d Know that the size and direction of the resultant force on an object affects its movement (**A** and **B**).

ACTIVITIES Forces and their effects

RESOURCES

A good reference book for teachers is *Forces – A Guide for Teachers*, published by the NCC and obtainable from NCC, Albion Wharf, 25 Skeldergate, York YO1 2XI

Concepts

The concepts involved in the study of forces are:

- A force can start an object moving.
- A force can make an object move faster.
- A force can make an object slow down.
- A force can stop a moving object.
- A force can change the direction in which an object is moving.
- A force can change the shape of an object.

MEASURING FORCE

RESOURCES A Newton meter

Force is measured in newtons on a Newton meter. A Newton meter (forcemeter) is a spring balance, calibrated in newtons (not kilograms).

Explain that forces can be measured using a forcemeter. Exhibit a forcemeter and demonstrate how it is used. Explain that scientists require a common unit for measuring forces so that tests can be fair wherever the scientist is working. Explain that force is measured in newtons.

Practical Give the children an opportunity to use the forcemeter to measure forces. Ask them to record their results and compare with results of others to check the accuracy of their measurements.

 FORCES

RESOURCES A large selection of different objects that move or can be moved in different ways

Class or group discussion Explain to the class that forces can act upon other forces of different sizes and directions. That a force can make an object change its speed or direction. Choose several of the objects you have collected and ask the children to think about the forces acting upon them. Discuss what they say and correct their ideas where appropriate. The *Forces – A Guide for Teachers* booklet from the NCC is a good reference for teachers undertaking this activity.

Practical Ask the children to choose several of the objects and to explore how they work or move and to try to identify the forces acting upon them. The children should record their findings so that groups can compare results.

ASSESSMENT **4.3c** Understand that forces can affect the position, movement and shape of an object.
4.4c Know that more than one force can act on an object and that forces can act in different directions.

 SPEED AND DIRECTION

RESOURCES A collection of objects that move in different ways. A suitable collection could include: a boat with a sail that moves across water, a toy car, a marble, a football or netball, a 'pull-along' toy

Class or group discussion Explain to the class that the size and direction of a force can affect the movement of an object. Choose several objects to explain exactly what you mean. Two different sized balls are probably the easiest way of explaining the principles involved but another example should be given that may not be so easy to understand, for example, a seed falling from a tree where the pull of gravity causes it to fall towards the ground with increasing speed.

Practical Ask the children to choose several different objects and explore the way that force can affect speed and direction. Record their findings so that they can compare results with other groups.

The children should be developing an understanding that the greater the force the greater is the acceleration of the object.

As far as is possible their investigations should be related to everyday situations such as road safety, cycling and sailing.

LEVERS AND FORCES

RESOURCES

A collection of common levers, such as spoons and screwdrivers of different lengths and scissors of different sizes. A set of cans with tight fitting, non-screw lids

Practical

Children can begin their work on levers by placing a ruler under a piece of newspaper, card, heavy cloth or a place mat on the edge of their desk. If they press their finger on the end of the ruler they can feel how easy or difficult it is to lift the material. Change the length of ruler placed under the material. What is the best position for the ruler so that the material can be lifted easily? Explain that the ruler is called a lever.

The children should be developing an understanding that the amount of force required to move an object depends on the length of the lever.

Practical

Give the children a can with its lid fixed firmly. Choose a set of objects and explore which is best for lifting the lid off the can. Children should record their results and use them to help them explain their investigation.

ASSESSMENT

4.5d Know that the size and direction of the resultant force on an object affects its movement.

INVESTIGATION   Forces and their effects

"What is friction, and is it always the same?"

Starting point
Picture of ice skaters and a discussion about sliding on the playground.

Observing and asking questions
The children should be encouraged to express their ideas about why skaters find it easy to glide across ice. Also about why it is easier to slide on the playground in icy weather. Discussion could also centre around the design of grips on shoe soles, and certain surfaces which are made to be anti-slip. Reference could also be made to roads and stopping distances for cars.

The children need to know that forces can be measured using a forcemeter. The children should also have undertaken Activity 1 AT 4(iii) Key Stage 2 (page 119) before they carry out any of the following studies. Encourage the children to look closely at the bricks and materials that are on display. If the bricks are going to be pulled across these materials what do the children think will happen?

Children may ask
Is friction the same on every surface?
Does the weight of the brick affect the amount of force needed to move it?
Does a slope affect the amount of force needed to move a brick?

Predicting and hypothesising
Taking into account the points that they discussed about differences in the ease with which the bricks move across the different materials, which

Resources
- bricks of different mass
- forcemeter
- selection of materials with different surfaces
- string

materials will allow the bricks to pass over them more easily? Consider the materials alone, not the bricks and materials together, as some children will find it difficult to look at more than one variable at once. Encourage children to explain their reasons in a scientific way. They may well begin to use the terms force and friction. Record the children's ideas to refer to at the end of the investigation.

Designing and planning the investigation

Before carrying out the investigation the children as a class could put together ideas about how they might investigate the forces needed for a brick to move over a material.

Ensure materials and equipment are on display to allow children to choose for themselves.

Some groups will need a considerable amount of teacher input, whilst others will work quite independently. The teacher should be encouraging the children to work as independently as possible in planning, setting out and carrying out their studies.

The more able groups could be asked to investigate what affects how a brick moves across different materials. This will involve them deciding for themselves what they are to explore and which variables they are going to control. They may wish to look at the mass of the brick over different materials, the amount of surface in contact with the material, or how much a slope affects the force needed.

Material type In this investigation the brick used will be the same for each test but the material under the brick must be different for each test. The children will attach the forcemeter and begin to pull. Once the brick moves the force on the meter will be recorded.

Brick mass This investigation will require bricks of different mass. The material under the bricks will be kept constant for each test.

Slope This investigation will require the brick and material to be kept constant, but the slope of the material to be changed for each test. It would be best for children to measure the angle of the slope for each investigation.

Recording

Before the children begin their investigation they should understand clearly what they are setting out to investigate, what they need to record, and how. Whilst undertaking the investigation, their recordings can be in rough and can be copied out clearly at a later date.

The children should be deciding on the most appropriate method of recording their results so that others fully understand.

Drawing conclusions

Ensure that the children's findings are based upon their investigation and not on their initial predictions. Is there a difference and can they see reasons for these differences? With their findings children could now explore ways of reducing the force required to move a brick across a wooden surface. This will give the children the opportunity to investigate the effects of using different lubricants to reduce frictional forces.

Wherever possible link the children's work to real life situations. For example, the Egyptians reduced frictional forces on the stone blocks used to build the pyramids by placing rollers under them.

Attainment Target 4 Physical processes

Key Stage 1 strand (iv)
Light and sound

PROGRAMME OF STUDY

A Sounds in the environment

Pupils should have the opportunity to experience the range of sounds in their immediate environment and to find out about their causes and uses.

B Echoes and the reflection of sound

They should experience the production of echoes resulting from the reflection of sound from distant surfaces.

C Making sounds

They should explore how to make and experience sounds by speaking and singing, striking, plucking, shaking, scraping and blowing, using familiar objects and simple musical instruments from a variety of cultural traditions.

D Light, shadow and reflection

They should have opportunities to explore light sources and the effects related to shadow, reflection and colour.

STATEMENTS OF ATTAINMENT

4.1c Know about the simple properties of sound and light (**B**, **C** and **D**).

4.2d Know that light passes through some materials and that when it does not shadows may be formed (**D**).

4.3d Know that light and sound can be reflected (**B** and **D**).

4.5f Know that sound is produced by a vibrating object and travels as a wave.

ACTIVITIES Light and sound

MUSICAL INSTRUMENTS

RESOURCES A collection of musical instruments. Explain how sound is produced from each instrument. There should be a wide range of instruments to ensure that the children are introduced to the fact that sound can be produced by plucking, striking, stroking, banging

Practical Give the children an opportunity to use the instruments and explore the sound they make. Can they change the sounds they make? How? Can they change the volume of sound created? How?

Children should be beginning to develop the idea that sound is made when something vibrates. They may be able to observe this in stringed instruments and can certainly feel the vibration in most instruments. They should understand that in order to set up a vibration an object has to be struck in some way.

Children should be developing an understanding that volume can be controlled not only by the force with which an object is struck but by the addition of a 'box' that amplifies the sound.

Once the children are aware of the key elements involved in creating and controlling sound they can set about designing and making their own instruments in design and technology. They can then work in music to compose and perform their own tunes.

ASSESSMENT
4.1c Know about the simple properties of sound and light.
The children are also developing an understanding in:
4.3d Know that light and sound can be reflected.
4.5f Know that sound is produced by a vibrating object and travels as a wave.

REFLECTING LIGHT AND SOUND

RESOURCES
A number of reflective plates (not glass), tubes of various lengths, card of different sizes and thicknesses, black and white card and sugar paper, Blu-Tack, Sellotape, some photographs to use as an image if required and any other useful materials you have to hand. If children require more tubes than you have encourage them to make their own

Class or group demonstration
Demonstrate how light, an image and sound can travel in a straight line to reach our eyes or ears. Use a long tube to demonstrate sound, a torch for light and yourself for the image.

Practical
Ask the children if they can get sound, light, or an image to go in a crooked line or an arc. Can they bend light, sound or an image?

Go through the resources that are available as these may give them some ideas of how to start. Discuss their plans so that they are clear about what they are going to do before they begin.

Encourage children who may come up with original ideas and only encourage them to try another method if you can see that they will have great problems leading to frustration.

Recording
The recording is more likely to be the final solution they devise and they should not be forced to write up and record everything they have done because it is a science lesson, or in order to simply acquire product for assessment files.

ASSESSMENT
4.3d Know that light and sound can be reflected.

LIGHT BEAMS

RESOURCES
Prisms, narrow-beam torches. Reference materials on light and rainbows

Practical Set up a display in the classroom with a series of questions which invite the children to explore:

1. The effect of shining a light on a prism. What has been created? Have they seen anything like it before?
2. A selection of materials, some of which let light pass through and some that do not. Encourage the children to combine the materials and observe the effects of shining light onto the different materials. Which materials let light shine through? Which materials reflect light? Which materials when combined stop the light waves passing through?
3. A selection of light filters and the effect of light shining through them. Which colours can they create by combining the filters?
4. The casting of shadows onto a white sheet of card. Be careful if using a lamp as children could get burnt.

Find time to talk to the whole class about what they have discovered at the display. When discussing their observations encourage them to use the correct language and to describe things in scientific terms whenever possible.

ASSESSMENT **4.2d** Know that light passes through some materials and that when it does not shadows may be formed.
And aspects of:
4.1c Know about the simple properties of sound and light.

INVESTIGATION Light and sound

"Are all shadows the same size?"

Starting point
A visit from a puppet theatre.

Observing and asking questions
The children can play at making shadows on the wall using their hands or by making their own puppets. The children could be encouraged to put on a 'shadow' puppet show of their own.

Resources
• a variety of torches
• card
• tape measures
• white sheet

Children may ask
What effect does the distance of the light source from the puppet have on the shadow?

Predicting and hypothesising
Encourage the children to make a prediction about what they think will happen and why.

Designing and planning the investigation
The same puppet should be used for each test and the distance between the light source and the puppet must be altered. The children will need to decide what distances they will investigate – and also how they will measure the resulting shadow.

Recording
This could be done through straightforward observation and a literal account of what happened.

Drawing conclusions
The emphasis here has to be on the teacher picking up the results of the investigation and talking to the children about why the shadow got bigger the nearer the light source was to the puppet.

Attainment Target 4 Physical processes

Key Stage 2 strand (iv)
Light and sound

PROGRAMME OF STUDY

A Sounds and how they are made

Pupils should learn that sounds are heard because they travel to the ear and that they can do so via a variety of materials. They should learn that sounds are made when objects vibrate, and investigate how sounds are changed in pitch, loudness and timbre, by changing the characteristics of the vibrating objects, for example, by changing length, tension, thickness of material of the vibrating object or the way it is made to vibrate, as exemplified by using musical instruments. They should be aware of the obtrusive nature of some sounds in the environment.

B The speed and reflection of light and sound

They should learn about the reflection of both light and sound and relate this to everyday effects (mirrors, echoes). Pupils should learn that light travels faster than sound by considering natural events such as thunderstorms.

C Light and shadow

They should explore the effects produced by shining lights through such objects as lenses, colour filters, water, prisms.

They should also investigate the formation of shadows and represent in drawings their ideas about how light varies in terms of brightness, colour and shade.

STATEMENTS OF ATTAINMENT

4.2d Know that light passes through some materials and that when it does not shadows may be formed (**C**).

4.3d Know that light and sound can be reflected (**B**).

4.4d Know that light travels faster than sound (**B**).

4.5e Understand how the reflection of light enables objects to be seen (**B**).

4.5f Know that sound is produced by a vibrating object and travels as a wave (**A**).

ACTIVITIES Light and sound

MAKING MUSICAL INSTRUMENTS

RESOURCES A collection of musical instruments

Class or group discussion

The teacher should recap on work covered previously (see KS1, Activity 2, page 123), a summary of which follows:

1. Sound travels through the air as waves that reach our ears.
2. Some materials are better than others for making sounds.
3. Sound waves can be reflected.
4. Volume can be increased with the addition of a 'sound box'.
5. Sounds can be made by making materials vibrate by plucking, stroking, banging.
6. Some materials absorb sound – as in a soundproof room. Many uncarpeted classrooms are creating the sense of more noise than the children are actually making.

Class or group demonstration

Children can find it difficult to imagine sound waves travelling through the air and it will help their understanding if the waves could be made 'visible'. If you take a tuning fork and strike it against a hard object and then lower it slowly into a bowl of water the children will be able to see the waves spreading out from the vibrating fork.

Practical

Children should explore the ways in which they can control pitch, loudness and timbre by changing the characteristics of the vibrating object or in the way it is made to vibrate. This can be undertaken with a collection of musical instruments or ones they have made themselves in design and technology.

Some children may find it easier to understand the principles involved if they make a simple plucking instrument by stretching an elastic band across a cardboard box. They will need some strips of wood to make bridges for the strings to rest on and to cut a hole in the box to create their own 'sound box'.

ASSESSMENT

4.3d Know that light and sound can be reflected.
4.5f Know that sound is produced by a vibrating object and travels as a wave.

REFLECTION OF LIGHT

RESOURCES

Prisms, lenses, coloured filters, water tray, bottles (empty and full of water), reflective mirrors (not glass), torches, spare batteries and a collection of good reference books on light

Class discussion

Teacher should recap on work undertaken at Key Stage 1.

Practical

The children should be given the opportunity to explore the effect of light shining on or through lenses, mirrors, prisms. They should record their findings carefully so that they can compare the different effects on light of shining it on the different objects.

Explain to the children that it is the light being reflected off objects that enables objects to be seen. This is a very difficult concept for young children to understand and may be remembered but not understood. Primary teachers should not be concerned as this will be picked up and developed further at Key Stage 3.

It is very rare that children experience conditions of total darkness. On some visits to caves the guide will cut out all the lights to allow the children to experience total darkness.

Research

Give the children time to research into how rainbows are made, what prisms do to light, how light is made and how it travels to create images in our eyes. This work could be undertaken in an English lesson by using the jigsaw approach where a group of children research an area that they know little about, pool their findings and then record what they think are the most important aspects of what they have discovered. Allow time for the children to report back and help to clarify the more difficult aspects of the way in which light travels.

Practical Once the children have gained a greater understanding that light is made up of the colours of the rainbow they can set about recreating the exact colours themselves. Once they are capable of recreating the colours of the rainbow they can make a colour wheel that can be spun and the children can observe how the colours combine to create white (see Figure 6). A reversal of what happens when they see a rainbow being created.

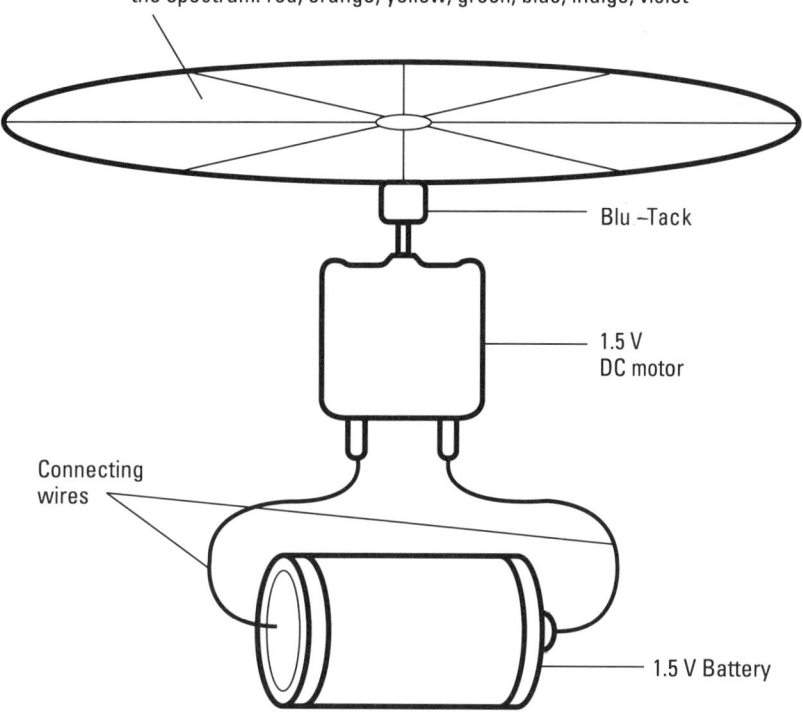

Card disc-children experiment using primary colours and colours of the spectrum: red, orange, yellow, green, blue, indigo, violet

Blu –Tack

1.5 V DC motor

Connecting wires

1.5 V Battery

Figure 6 Making a colour wheel

3 LIGHT AND SOUND

Aim To further understanding that light waves travel faster than sound waves.

Class discussion Recap on work covered previously and then explain that light waves travel faster than sound waves. Explain that light waves create what we see and sound waves create what we hear. Can the children think of examples of when we see things before we hear them?

- a plane travelling towards us;
- the 'bang' of a starter pistol;
- the light in lightning is seen before we hear the thunder;
- a golfer hitting a golf ball.

ASSESSMENT **4.4d** Know that light travels faster than sound.

INVESTIGATION Light and sound

"How does a guitarist change the sound his instrument makes?"

Resources
- elastic bands
- shoe boxes
- wooden blocks
- old guitar
- large cardboard boxes

Starting point
Watching a video of a Spanish guitarist.

Observing and asking questions
After watching the video of the guitarist discuss with the children how the sounds are made, how the pitch is changed and how the volume of the guitar is increased.

Children may ask
Does thickness of the string affect the sound?
Does length of the string affect the sound?
Does the size of the soundbox affect the volume?

Predicting and hypothesising
Encourage the children to make a prediction before carrying out their investigations and to give reasons for their predictions.

Designing and planning the investigation
Some children will undertake an investigation into the effect of changing the length of the elastic band (see Activity 1, page 125). This can be done by placing blocks of wood at different lengths underneath the elastic band.

Some children will realise that the thickness of the band affects the pitch and may be able to investigate both thickness and length together and observe their effect on pitch.

Some will investigate volume by placing the same thickness of band over different sized boxes.

Some may realise that the force used to pluck the band affects volume and will look at ways of controlling the force to undertake a fair investigation.

Recording
This investigation will be difficult to record in any quantifiable way. However there is a sensor that can be attached to the BBC Microcomputer called the Panthera Measuring Box (obtainable from 5, Cedar Avenue, Beeston, Nottingham NG9 2HA) which will give a screen display and print-out of the volume of sound created. The children will need to trust their ears to give judgement on pitch.

Drawing conclusions
Children should be encouraged to formulate their findings in scientific terms. *The pitch depends upon the length of the elastic band. The higher the pitch the shorter the elastic band. The thicker the elastic band the lower the pitch. The volume created depends upon the size of the soundbox. The bigger the soundbox the greater the volume. The bigger the soundbox the greater the sound when struck or plucked with the same force.*

The children's conclusions should be based upon their investigations which they should use to explain their findings.

Children could move on from this investigation to Activity 1 where they make their own instruments on which to compose their own music.

Attainment Target 4 Physical processes

Key Stage 1 strand (v)
The Earth's place in the Universe

PROGRAMME OF STUDY	**A Seasonal change**

A Seasonal change

Pupils should observe closely the local natural environment to detect seasonal changes, including length of daylight, weather and changes in plants and animals and relate these to the passage of time.

B The changing appearance of Sun and Moon

They should observe, over a period of time, the length of daylight, the position of the Sun, and when possible the position of the Moon in the sky and its changing appearance.

STATEMENTS OF ATTAINMENT

4.1d Be able to describe the apparent movement of the Sun across the sky (**B**).

4.2e Know that the Earth, Sun and Moon are separate spherical bodies (**B**).

4.3e Know that the appearance of the Moon and the altitude of the Sun change in a regular and predictable manner (**B**).

ACTIVITIES The Earth's place in the Universe

THE MOON

RESOURCES Model spheres showing how the Moon revolves around Earth

Class or group discussion

Observing the Moon.

Children could be encouraged to look whilst at home but may occasionally see it in the morning.

What is the Moon?
When can we see it?
Does it always look the same?

The children could keep a diary showing the shape of the Moon at different times in the month.

Class or group demonstration

The Moon revolves around the Earth in one month. This could be demonstrated and discussed using model spheres.

<table>
<tr><td>ASSESSMENT</td><td>The activity has scope to add to a child's overall understanding of:
4.2e Know that the Earth, Sun and Moon are separate spherical bodies.</td></tr>
</table>

A DIARY OF THE SEASONS

RESOURCES | A diary to keep a record of the changes they observe taking place

This activity introduces the children to the Programme of Study which states that: 'Pupils should observe closely the local natural environment to detect seasonal changes . . . including changes in plants and animals and relate these to the passage of time.'

Practical | The children could keep a monthly diary to record their observations of minibeasts around the school grounds. Questions could be asked as follows:

1. On an Autumn day. What changes are taking place to the leaves?
2. On a Spring day. Where can you see plants showing signs of growth and new life?
3. On a Summer day. How many different creatures can be found in the grass?

<table>
<tr><td>ASSESSMENT</td><td>There are no Statements of Attainment which could be assessed through this activity but the experience will help create a foundation of understanding needed at Key Stage 2 for:
4.3e Know that the appearance of the Moon and the altitude of the Sun change in a regular and predictable manner.</td></tr>
</table>

LENGTH OF DAYLIGHT

RESOURCES | A collection of reference and reading books that refer to the seasons and day and night

Class or group discussion | At which time in the year does it become dark early?
Which season of the year gives us light evenings?
Where do you play after school in the Summer and in the Winter?
Do the days get longer and shorter?
What makes the Summer days seem longer and the Winter days seem shorter?

4 KEEPING A WEATHER CHART

RESOURCES | A weather chart and equipment for recording the temperature and wind direction

Practical | Explain clearly how to read the temperature and wind direction using the equipment available. Show the children how to complete the weather chart.

What is the weather like today?

Is today warmer or colder than yesterday? How do you know?

Demonstration | The children should be taught how to use a thermometer. (It is possible that here the teacher could build upon the introduction to the thermometer described in AT4(ii) Key Stage 1, Activity 1, page 107).

Daily recording will help the children to understand the meaning of the scale.

Practical Are sunny days always the warmest?

The children could keep a class record to show cold, warm and hot days.

Explore whether certain parts of the school playground are warmer than others. Which playtime is usually the warmest on a sunny day?

5 'MY DAY'

RESOURCES Charts on which to set down what each child does on a normal day

Class or group discussion What things do you do in a day? (Examples might be: go to school, play, sleep.)

The children could show with pictures or charts what they have done each day for a week. Then discussion could invite the class to think how to show the times when it is night and the hours of daylight.

Are your days different in the Summer or Winter?

Set up records of their day and compare the activities they undertake, when they go to bed, when they get up. What do they do when it is dark? What do they do when it is light? What difference does it make to the activities they undertake when it is warm?

ASSESSMENT Activities 3 to 5 combined offer opportunities to assess:
4.1d Be able to describe the apparent movement of the Sun across the sky.

INVESTIGATION

"What are shadows?"

Resources
- variety of objects that will create shadows
- rulers
- tape measures
- camera (not essential)
- video recorder (not essential)
- chalk

The Earth's place in the Universe

Starting point
A walk around the school on a sunny day.

Observing and asking questions
Initially look for all the shadows that can be found on your walk. Discuss with the children why shadows appear. The children could be encouraged to put out a variety of objects to see what shapes of shadows they cause. Some children may suggest that the shadows will be different at different times during the day. This should lead to a variety of questions which could be investigated.

Children may ask
Do shadows get longer during the day?
Does the position of shadows change during the day?

Predicting and hypothesising
The children could predict what will happen in their investigations and give reasons. It may be possible for children to give reasons for their prediction which reflects quite a high level of knowledge. Some children may be aware that the position of the shadows will change because of the apparent movement of the Sun across the sky. It is unlikely at this stage that they will be aware that it is the Earth moving in relation to the Sun that affects the size and position of the shadows.

Designing and planning the investigation
The children will need to decide which places around school will be best for

investigating shadows. They will also need to decide over what time period they are going to observe them. If they are conscious that shadows in the Summer are different to those in Winter, then it may well be a very long-term investigation.

A good scientist makes sure that the investigation is a fair one.

If the children are unsure what is meant by fair investigation then you may need to explain that they must take care to ensure that they always do the same things at the same time each day.

Length and position of shadow Both these investigations can be carried out at the same time. Simply, the children have to decide at what intervals they are going to record the length and position of the shadow.

Recording
The recording could be done as a group or as a whole class over a long time period. If the whole class undertakes the investigation then it can be easier for the teacher to manage but more difficult to undertake individual assessment.

This investigation allows the teacher to observe children choosing and using measuring and recording equipment. Teachers should ensure that the children have opportunities to choose their equipment to measure and record with.

Drawing conclusions
This aspect of investigation is probably the most important as it will be at this stage that the teacher will need to try and draw out from the children what happened and why.

Hopefully the children will have begun to realise that the Earth is moving in relation to the Sun and therefore our position on the Earth changes in relation to the Sun throughout the day.

This activity is an excellent introduction to the idea of the Earth's position in the Solar System and how it moves around the Sun every year.

Attainment Target 4 Physical processes

Key Stage 2 strand (v)
The Earth's place in the Universe

PROGRAMME OF STUDY

A Following the path of the Sun

Pupils should track the path of the Sun using safe procedures such as a shadow stick or sundial.

B Watching the night sky

They should study, using direct observations where possible, the night sky including the position and appearance of bright planets and the Moon.

C The movement of the Earth, Sun and Moon

They should learn about the motions of the Earth, Moon and Sun in order to explain day and night, day length, year length, phases of the Moon, eclipses and the seasons.

D The Solar System

They should be introduced to the order and general movements of the planets around the Sun.

STATEMENTS OF ATTAINMENT

4.2e Know that the Earth, Sun and Moon are separate spherical bodies (**A,B,C** and **D**).

4.3e Know that the appearance of the Moon and the altitude of the Sun change in a regular and predictable manner (**A, B** and **C**).

4.4e Be able to explain day and night, day length and year length in terms of the movements of the Earth around the Sun (**A** and **C**).

4.5g Be able to describe the motion of planets in the Solar System (**D**).

ACTIVITIES The Earth's place in the Universe

THE PLANETS

RESOURCES   A selection of good reference books on the planets and the Universe

Class or group discussion   The teacher will need to have background knowledge to be able to talk to the class about the positions and movement of the planets around the Sun.

Practical   The children could use reference material to draw diagrams or to make a model of the Solar System. Older children enjoy finding out as much as they can about the planets and the Universe.

Research A scale representation (1cm=1 million miles) could be made on the playground or school field.

Distances in millions of miles of the planets from the Sun are as follows: Mercury, 36; Venus, 67; Earth, 93; Mars, 142; Jupiter, 484; Saturn, 887; Uranus, 1785; Neptune, 2797; Pluto, 3665.

How long does it take for the Earth to complete one orbit of the Sun? How long is a 'year' on other planets?

ASSESSMENT **4.5g** Be able to describe the motion of the planets in the Solar System.

2 SUNDIALS

RESOURCES A simple object, for example a length of stick, in the ground to cast a shadow

Discussion Sundials made by the children can be discussed.

Why does the shadow move round the sundial?
When do we get long and short shadows?
What is the shortest period of time which is possible to record on a home-made sundial?

Practical Look at how shadows on the classroom wall move throughout the day. Could these shadows be used to give an approximate time of day? How?

The children could also study the sundial designs which are popular in gardens and on buildings and evaluate which is the easiest to use and interpret.

3 DAY AND NIGHT

RESOURCES A torch, a large ball to represent the Earth and a small ball to represent the Moon

Class or group discussion Explain to the children how the Earth rotates around the Sun, and the Moon around the Earth.

Use the torch to represent the Sun to explain how night and day are created by this movement.

What is happening to the Earth as it moves in space around the Sun?
How long does the Earth take to complete one spin?
How long does it take for the Moon to travel around the Earth?
Which parts of the Earth are closest to the warmth of the Sun?

ASSESSMENT **4.2e** Know that the Earth, Sun and Moon are separate spherical bodies.
4.3e Know that the appearance of the Moon and the altitude of the Sun change in a regular and predictable manner.
4.4e Be able to explain day and night, day length and year length in terms of the movements of the Earth around the Sun.

4 THE NIGHT SKY

RESOURCES A selection of good reference books on stars and constellations. A number of boxes painted black inside so that constellation patterns can be made

Children can research the different star patterns and their names.

Practical
The patterns of the constellations can be made by putting holes in black card, in the shape of the constellation, and holding the card up to the light. Similar holes could be made in an old umbrella to create the effect of a night sky 'canopy'.

Can you find the Plough, Orion or the Pole Star?

Class or group discussion
Explain to the children that a star is a sun and is visible in the night sky. Explain why we cannot see stars during the day.

It is more difficult to view the stars in a city because of the glare of lights. The best place to view a night sky is either on a school, residential, field trip or in a planetarium. Children can then identify the star patterns they had created in the classroom.

Bright planets such as Venus and Mars are easy to locate at times when they are close to the dawn/dusk horizon. The astronomy section of most national newspapers often indicates when and in which direction these planets can be seen.

Many reference books show stunning photographs from the Space Programme and could be used to promote discussion.

The Observer Book of Astronomy by Patrick Moore, published by Frederick Warne and Co., is an excellent reference. *The Daily Telegraph Map of the Night Sky* can also be recommended.

ASSESSMENT
This activity offers no opportunity for assessments to be undertaken but addresses the Programme of Study which states that children should be given the opportunity to observe the night sky.
The activity does however add to the children's overall understanding of the Universe.

THE SEASONS

RESOURCES
A torch and a sphere to represent the Earth

It is more realistic if the children can make a papier mâché Earth with the Equator clearly marked on it.

Class or group discussion
Use the model to explain to the children how the Earth is tilted and rotates. Use the torch to represent the Sun and show how the Earth rotates around the Sun. Explain how this determines day and night and length of day and the year.

Why are some parts of the World colder or warmer than others?
How does the fact that the Earth's axis is tilted at an angle (66°) create the seasons?

Practical
The children could use the torch and model to explain to a friend how this tilt creates Summer and Winter.

How many hours of daylight and darkness do we have at different times of the year?
How can we find out? (Reference sources such as lighting up times in diaries and newspapers could be mentioned.)

ASSESSMENT
4.4e Be able to explain day and night, day length and year length in terms of the movement of the Earth around the Sun.

INVESTIGATION

"Is the Moon always the same in the night sky?"

The Earth's place in the Universe

Starting point
Observing the Moon.

Observing and asking questions
The teacher may need some information to help explain the changing shape of the Moon as it revolves around the Earth each month.

The Moon, our nearest neighbour in space, is slightly less than 239 000 miles away. The new moon, which is almost in line with the Sun, is invisible because it reflects none of the Sun's light. A few nights later it can be seen as a thin crescent. Each night, rising later, it changes to half Moon (first quarter), then gibbous, and once the whole of the hemisphere is turned towards the Earth it becomes the full Moon.

Binoculars provide impressive views of the surface of the Moon. There are times in the school day when the Moon can be seen clearly but it is likely that opportunities to make sketches may have to be undertaken from home.

Children should be encouraged to find answers to questions such as the following.

Can we always see the Moon?
Is it always there?
What does the surface look like?

The children could compare their observations with photographs and perhaps videos of the Moon Landings.

Research could be carried out to assemble facts about the Moon.
What is the Moon?
How is it similar/different to planets in the Solar System?
What is the pattern of its movement?
Which other planets have moons?

Predicting and hypothesising
Encourage the children to see if they can predict what the shape of the Moon will be.

Designing and planning the investigation
The children's investigation could set out to see if there is a regular pattern to the changes which we see.

Sketches of the Moon's shape could be kept over a period of time in a Moon diary. From their results the children could decide if we get a full Moon at the same time each month.

Printed diaries could be used to support first-hand experience.

The children's observations and curiosity should be supported by encouraging study skills and the use of reference books.

Recording
Consideration needs to be made as to how the children could display their observations and explain them to friends in class.

A simple model could be made to show how the Moon is illuminated by the Sun.

Drawing conclusions
From the results the children have recorded they could see how the time spans, which produce day, month and year, have been measured using the movements of the Earth, Sun and Moon.

Work could also be done on how the Moon affects our tides.

6

DEVELOPING ATTAINMENT TARGET 1

Planning and assessing Attainment Target 1

THE IMPORTANCE OF ATTAINMENT TARGET 1

Primary teachers should be aware that the overall aim of Attainment Target 1 is:

> *That children are working towards developing the scientific knowledge and skills to enable them to plan, undertake and evaluate independently scientific investigations*

This should be viewed as the overall achievement that schools are working towards.

Attainment Target 1 is of special importance in science in that it:

– demands that children undertake scientific investigation;
– aims to develop scientific skills in addition to scientific knowledge;
– aims to develop in young children a greater awareness of how scientists work;
– may well have significant weighting in overall assessments against other attainment targets.

Science skills, as outlined in Attainment Target 1, will only be achieved over a period of time and not all children will become totally confident and proficient during their primary education. For children working towards this achievement it is important that primary school staff plan as a team to identify the progression within Attainment Target 1 and from this to plan appropriate investigations.

Our advice is for schools to draw up, over a period of time, a bank of investigations for use throughout Key Stages 1 and 2. The investigations should be set within a variety of contexts and cover work related to all the attainment targets. This support material provides a bank of experiments, which teachers can use in various ways, for example, as starting points, adapted or used as provided.

When putting together a bank of experiments it should be recognised that certain parts of a strand are more successfully undertaken through investigation.

THE CHALLENGE FOR TEACHERS

IDENTIFYING THE SCIENTIFIC SKILLS

To identify and understand within the Programmes of Study the range of skills to be developed within the scientific investigation.

PLANNING FOR PROGRESSION

To set down the progression through Key Stages 1 and 2 in terms of skills.

PLANNING SCIENTIFIC INVESTIGATIONS

To plan carefully so that before each investigation is undertaken it is:

– clear what is being taught;
– known what can be assessed.

This can only be achieved by planning jointly from the Programmes of Study and the Statements of Attainment. This is crucial if investigations are to stay on task and teachers are not to reach the end of an investigation saying there was nothing that could be assessed.

CLASSROOM MANAGEMENT

Managing the investigations within a class of thirty or more children of a wide ability range.

Curriculum Management and Classroom Practice in Primary Schools, Alexander, Rose and Woodhead, DES 1992, para 97, emphasises a strong view shared with HMI that, 'If group work is to be employed, each teaching session should focus on a manageable number of group and learning activities.'

UNDERTAKING ASSESSMENTS

Assessing statements of attainment for Attainment Target 1 within an investigation.

Undertaking assessments within the environment of the busy classroom and where individuals have to be assessed during group activities.

Teachers need a clear understanding of how to assess children whilst they undertake an investigation. Teachers need to decide what to look for before they introduce the investigation.

INCREASING CHILDREN'S INDEPENDENCE

Teachers need to plan the balance between teacher-directed investigations and those where children plan and undertake investigations independently. This is a tremendous challenge as younger children will require more teacher direction and the balance changes as children progress through the levels set down within the National Curriculum. With a wide range of ability at the upper end of primary school there will be similar investigations being undertaken but with a great difference in the amount of teacher direction required.

RAISING CHILDREN'S AWARENESS OF 'FAIR' TESTING

Teachers will need to be clear about the differences and will need to plan to allow for children to predict and hypothesise and to work with and control variables. Children's understanding should begin with the fact that investigations need to be 'fair' and progress to children identifying variables and how to control and change them.

TEACHER-ORGANISATION GUIDE

Throughout section 5 all the investigations have been set to a common format designed to help teachers plan their own investigations.

The investigations introduced are broad in that they have been written to a key stage. Teachers will therefore need to adapt them to match the ability of the children in each class.

To assist teachers in adapting the investigation, or writing their own, we have compiled questions which teachers will need to consider under each heading when planning their own investigation.

Investigation planner

ATTAINMENT TARGET _____

AT _____ KEY STAGE _____ INVESTIGATION_____

TEACHER QUESTION

Set the context for the investigation.

Ensure the question is open-ended to encourage the children to devise their own investigation. For example, 'Investigate what things are needed for seeds to grow.'

The question provides opportunities for children to have more control over what they want to investigate – light, water, soil or type of seed.

If the question is closed then the children can make no decisions on what to investigate. For example 'What type of soil is best for growing seedlings?' allows the children only to investigate the type of soil!

The investigation should allow some decisions to be made by the children regarding predicting, questioning and originating ideas.

RESOURCES

Decide on the equipment and materials which will be required. Have the children gained experience of using them or will you need to give guidance and instructions or an opportunity to use the equipment before they undertake the investigation?

It is important that children are given the opportunity to choose and use equipment and material if teachers are to assess at the higher levels.

What equipment will be to hand for children to choose? Have the children been trained to use the equipment safely and correctly? At level 4 and above children should be choosing equipment independently and using it correctly. Prior to level 4 the teacher can demonstrate how to use the equipment but once shown the children should be able to use it correctly themselves. Teachers need to decide to plan for opportunities for the children to select materials and equipment themselves or to restrict them in their choice.

STARTING POINT

It is important that teachers choose suitable starting points which will act as a stimulus to children undertaking the investigation or that places it within context of the children's experience.

OBSERVING AND ASKING QUESTIONS

Provide a variety of opportunities to promote discussion between the children, teacher and children, and children and other adults.

The amount of teacher guidance and input will vary considerably depending upon:

– The children's previous experience in undertaking investigations;
– The complexity of the investigation being undertaken;
– The age of the children;
– The general ability of the children;
– Assessments being undertaken. Teachers may have to sit with children and ask appropriate questions to give children the opportunity to predict in order to assess the different levels of ability within a group.

Teachers should use their professional judgement concerning the amount of input they give.

Give every opportunity to handle, observe and discuss the area being investigated.

At this stage teachers may need to increase the children's knowledge of the area to be investigated.

In writing the investigations due note was given to *Curriculum Organisation in Classroom Practice in Primary Schools: a discussion paper,* Rose and Woodhead, which calls upon teachers to match teaching style to work being undertaken. The main implications of the report for the teaching of scientific investigations are as follows.

- There are times when it is easier, more effective and time saving to gather the whole class together to give them information that they are all required to know.
- There are times when discussions might best be undertaken with a whole class.
- Children might undertake research, watch a demonstration, explore and discover, or have the teacher explain something to the whole class.

What is important is that the teacher matches the most appropriate approach to the task in hand.

CHILDREN MAY ASK

Teachers should encourage the children to pose their own questions that can be investigated.

They cannot ask questions if they have nothing on which to draw and so should have been provided with sufficient stimulus, knowledge and direction to have motivated them to pose sensible questions which can be investigated.

The questions may be verbalised or written down but often children undertake work having already asked themselves the questions. The child has independently selected two healthy plants and placed them in different conditions of light, for example.

ISOLATING ONE VARIABLE

Most investigations at Key Stage 1 and at Key Stage 2 should direct children to pose questions that allow them to investigate a single variable. When children are undertaking an investigation into growth of plants, they may plant seeds and place them around the classroom. Some plants thrive, some do not and some do not grow at all. At the end of the half-term the children may be aware that some seeds grow and others do not but have learnt little about the effects of water, light or soil type on the growth of a plant. It is at this stage that the teacher needs to ensure that the children are fully aware of what it is that they are to investigate and observe. It is for the teacher, through careful manipulation, to assist the children in isolating just one variable, for example, water, that affects the growth of plants.

Such an approach, as well as being more effective in developing children's scientific knowledge, will make the investigation easier for the teacher to manage and for the children to undertake. The children will need to be gathered together at regular intervals to explain their work to others so that they are aware of the way others undertake their investigations and gain greater knowledge of the effects of other variables on the growth of plants.

The more able children at Key Stage 2 should be given the opportunity to investigate more than one variable at a time.

PREDICTING AND HYPOTHESISING

Encourage children to predict what they think will happen in their investigation prior to undertaking it.

To assist teachers in their planning and carrying out their assessment it is helpful to see children as progressing through four stages at Key Stages 1 and 2 linked to the following National Curriculum levels:

141

Level 2 **Predicting and hypothesising**
Children make a prediction.

Children may say
I think the more stirs you give it the faster the sugar will disappear.

Level 3 **Predicting and hypothesising**
Children make a prediction based on an everyday experience.

Children may say
I think the more stirs you give it the faster the sugar dissolves because my Auntie always stirs her tea for a long time to dissolve the sugar quickly.

Level 4 **Predicting and hypothesising**
Children make a prediction based upon relevant prior knowledge.

Children may say
I think the sugar will dissolve more quickly if placed in hot liquid because when we made jelly it dissolved more quickly in hot water.

Level 5 **Predicting and hypothesising**
Children make a prediction based upon relevant scientific knowledge.

Children may say
I think when you add sugar to hot water it will dissolve more quickly because the particles move about more quickly when heated.

DESIGNING AND PLANNING THE INVESTIGATION

Children should be given the opportunity to plan and decide how they are going to undertake the investigation. However, younger and less able children may require more teacher input and guidance.

Once children have decided on their investigations they will need to:

– Identify how they can make their investigation 'fair'.
– Decide which variable they are going to change. If they are to investigate the effects of the quantity of water on the growth of plants then they will need to change the quantity of water given to each plant. Everything else in the investigation will need to be kept the same so that the children can see more clearly the results and the teacher can more easily manage the investigation.

Children should be considering the health and safety aspects and care of living things within the investigations they plan.

Teachers should not allow children to proceed further until they are clear exactly what is to be investigated and how the children will carry out their investigation.

LOOK BACK AT THE LIST OF RESOURCES

Are there any health and safety aspects that you need to bring to the children's attention?

Are there any aspects of care and consideration for plants, animals and/or others that need to be brought to the children's attention?

Consider the aspects of Information Technology that might be used to assist in the investigations. The following may be considered.

– Making an observational drawing using an art program;
– Presenting data in an appropriate format. For example, pie chart, line graph, scattergram;

– Monitoring changes in temperature or light over a period of time;
– Using a data handling program to interrogate and present data.

RECORDING

Children should make decisions about how they record their observations and in what form.

In order to make decisions children will need to have had experience in recording using:

– graphs;
– charts;
– data bases;
– diagrams;
– drawings;
– close observational drawings;
– notes.

The method children use for recording should communicate effectively to others the information they have gathered. Set the children the challenge of explaining their findings to others in the clearest possible way.

There should always be a purpose for children's recordings so that they can be used for interpretation and drawing conclusions.

DRAWING CONCLUSIONS

Children should be drawing conclusions based upon what they observed in their investigation.

Conclusions should be based upon the data collected or observations made within the children's investigation and not solely on their original prediction.

Children should comment as to whether their investigation was 'fair'. At higher levels children should be giving reasons for their comments and giving guidance on how the investigation could have been improved.

Children should be looking for patterns within the data they have collected.

At Level 3 and above, children should be reporting on what happened and giving reasons. At this level teachers should be looking for children to be quantifying their results. Instead of saying the plant grew 30 cm children can say the plant grew 2 cm every week.

Summary of skills in Attainment Target 1

	Strand (i) Ask questions, predict and hypothesise	Strand (ii) Observe, measure and manipulate variables	Strand (iii) Interpret their results and evaluate scientific evidence
Key Stage 1	The children should be: – sorting, grouping and describing materials and events; – using firsthand observations; – making predictions; – coming up with ideas to investigate; – raising questions; – using scientific knowledge; – increasing scientific knowledge.	The children should be: – using standard and non-standard measures; – 'fair testing'; – carefully observing investigations; – developing awareness of safety in undertaking investigations.	The children should be: – giving simple explanations for what happened in an investigation; – evaluating their investigations; – interpreting results; – recognising a fair test; – developing understanding of purposes of recording; – using appropriate recording; – drawing conclusions.
Key Stage 2	– making predictions based upon scientific knowledge; – using secondary sources to help formulate questions and to inform predictions; – using firsthand observation to help formulate questions and to inform predictions.	– using computers and sensors to capture, transmit and store information; – independently using equipment and materials; – quantifying variables; – identifying and controlling variables to be investigated (carrying out a fair test); – using IT to support investigations; – undertaking more precise investigations; – increasing understanding of safety and care in carrying out investigations.	– evaluating their own investigations and suggesting how they might be improved; – considering different interpretations of the evidence of their investigations; – interpreting data and evaluating them against the original question; – answering questions; – searching for patterns of data; – using IT to record and to help look for patterns in data.

Progression in Attainment Target 1

	Strand (i) Ask questions, predict and hypothesise	Strand (ii) Observe, measure and manipulate variables	Strand (iii) Interpret their results and evaluate scientific evidence
Level 1		Children observing in the context of a scientific investigation materials and events in their immediate environment using their senses. Being encouraged to communicate their observations either orally or through writing or drawings.	
Level 2	Children asking questions, suggesting ideas to investigate and making predictions. Teacher helping by providing open-ended questions to allow the children to formulate their own questions within the context set by the teacher. The children do not have to give reasons for their predictions.	Children making observations which are connected to questions or ideas that are being investigated. Beginning to use non-standard and standard measures in quantifying their observations. Observations at this stage could be in pictures, words or through dialogue.	Children drawing conclusions based upon observations made. Relating their conclusions to their original prediction or idea. The conclusions could be oral or written. There is a link between strands (i) and (iii) because the children need to compare what they saw with what they expected.
Level 3	Children asking questions which can be investigated and making predictions themselves. As at Level 2 the context can be set by the teacher. Children giving reasons for their predictions based upon everyday experience.	Children using a variety of measuring instruments to support observations in their investigations. Teacher supporting the children by helping them to choose appropriate measuring instruments and showing how to use them.	Children saying whether or not their investigation was fair. Explaining how the unfairness of the test may have affected the conclusion they have drawn. Not only reporting what happened but also saying why they think it happened.
Level 4	Children asking questions to investigate and making predictions. Giving reasons for their predictions which are based upon relevant prior knowledge. This relevant knowledge may be based upon work from previous science lessons or work from other curriculum areas.	Children setting up a fair test. Showing an awareness of the variables they are investigating. Independently selecting and correctly using appropriate materials and equipment.	Children looking for patterns in the data they have collected for their investigation. Attempting to link the patterns in the data to the original question, prediction or idea which they are investigating.
Level 5	Children asking questions to investigate and making predictions. They should be giving reasons for their predictions which are based upon scientific knowledge that reflects knowledge of science at Level 5 in the knowledge-based attainment targets.	Children handling a range of variables within an investigation, altering the range of the variable being investigated and selecting appropriate amounts for those variables which are being kept the same. At this stage the children should be working with continuous independent variables.	Looking for explanations for what has happened within their investigation. The teacher encouraging the children to look for more than one explanation for what happened within the investigation they have carried out.

Teacher's planning sheet

This planning sheet is provided to help teachers to focus on the various stages that children may work through when undertaking a scientific investigation. The order the children work through the stages will vary according to the investigation being undertaken.

The sheet is designed to assist teachers in keeping 'on track', in ensuring that the activity remains an investigation and helping to identify opportunities for undertaking assessments.

BEFORE THE INVESTIGATION

1. What do you want to find out in your investigation?

2. How will you undertake your investigation?

   – How will you make your investigation 'fair'?
   – What variables are you going to change?
   – Where will you undertake your investigation?
   – What equipment are you going to use?
   – Is your investigation safe?

3. What do you think will happen?

   – Can you give a reason for your prediction?

4. How will you record your findings?

5. Carry out the investigation.

AFTER THE INVESTIGATION

6. What happened in your investigation?

7. Was your investigation 'fair'?

8. Can you explain why you think it happened?

9. If you did the investigation again what would you do differently?

Appendix 1 Computer resources

The following list of equipment has been found to be the most suitable for use in primary schools. It is recommended that an old BBC computer system be dedicated to computer control so that equipment can be left connected and children can be given opportunities for writing simple procedures to control lights and switches during any spare time they may have. Computer control work will cover a great deal of the Programmes of Study in Information Technology, especially those under Applications and Effects and Measurement and Control .

EQUIPMENT

DELTRONICS PRIMARY 'CONTROL IT' INTERFACE BOX

This package includes several sensors but nowhere near enough, nor those most commonly in use. The kit does not contain any software with which to run the control system.
Deltronics
91 Heol-y-Parc, Cefneithin
Llanelli, Dyfed SA14 7DL
Tel: (0269) 843728

PRIMARY TECHNOLOGY: THE PLACE OF COMPUTER CONTROL

This is a very useful pack which includes the software *Contact* required to run the control equipment on a BBC computer, in addition to many ideas for activities to be carried out by primary children.
NCET Publications
Sir William Lyons Road, Science Park
Coventry CV4 7EZ
Tel: (0203) 416994

EXTRA OUTPUT AND INPUT LEADS (WITH SENSORS ATTACHED)

Output leads have the larger 3.5 mm jack plugs. The most useful outputs to order are those with lights and buzzers. Lego motors are advised instead of the cheaper alternatives for control work as they can easily be incorporated into Lego models made by the children. When using Lego motors order extra output leads with Lego motor attachments already fitted.

Input leads have the smaller 2.5 mm jack plugs. The most useful inputs to order in larger quantities are those with LDR (light sensors) and switches attached.

The equipment can be purchased from:
GC Products
The Bungalow
Northbeck, Scredington
Sleaford, Lincolnshire NG34 0AD
Tel: (0529) 306234 or (0529) 305613

BOX OF LEGO TECHNICS 2

– Additional Lego motors
– Lego motor speed controllers (gear blocks for use with motor sets 8700 and 880. Lego Part No. 872.)

The Lego equipment can be purchased from:
Technology Teaching Systems
Unit 4
Holmewood Fields Business Park
Park Road, Holmewood
Chesterfield S42 5UY
Tel: (0246) 850085

Appendix 2 Bibliography

Teachers may find the following books, listed by publisher, a useful reference source.

Classroom
Practice in Primary Schools

HMSO Books, London
Science; Key Stages 1, 2 and 3

Lincolnshire County Council, Lincoln
Science at Key Stages 1 and 2

Macdonald 5–13, London
Minibeasts
Ourselves
Trees
Seeds and Seedlings
Metals–Stages 1 and 2
Metals–Background Information
Structures and Forces

Macdonald Educational, London
An Early Start to Science
An Early Start to Nature

National Curriculum Council, York
Forces – A Guide for Teachers
Electricity and Magnetism

Science Horizons
Our Home
Keeping Our Home Warm
Materials In The Home

Science Now
Plastics A Plenty
Metals and Corrosion
Chemicals On The Farm
Electricity
Inheritance
Biotechnology
The Body In Action

Technology Teaching Systems
Electricity Book 1
Electricity Book 2

The Association for Astronomy
Earth and Space

The following books, listed by publisher, provide useful resources for children.

Dorling Kindersley, London
My Science Book of:
Electricity
Magnets
Light
Colour
Sound

Eyewitness Visual Dictionaries:
The Visual dictionary of Plants
The Visual Dictionary of Animals

Eyewitness Guides:
Fossil
Rock and Mineral
The Earth and How It Works
Fish
Seashore
Pond and River
Mammal
Insect
Butterfly and Moth
Shell
Plant
Bird
Skeleton
Amazing Spiders
Amazing Frogs and Toads

Oxford University Press, Oxford
The Young Scientist Investigates:
Seeds and Seedlings
Flowers
Trees
The Human Body
Air
Water
Rocks and Soil
Everyday Chemicals
Materials
Into Science – Weather

Usborne, London
Planet Earth
Children's Encyclopedia of Prehistoric Life

Thomas Nelson, Walton-on-Thames
Science – Start Here:
Making it Move
Making Changes

HarperCollins, London
Key Stage 1 Toys

Ginn Science, Aylesbury
Sound
Wood and Paper

Blackie Academic and Professional, Glasgow
Sound

Wayland, Hove
Water

Appendix 3 Blank Programme for Science

A TWO-YEAR CYCLE (YEARS A AND B) FOR KEY STAGE 1 AND KEY STAGE 2 WHICH SHOWS:

- Each strand for Attainment Targets 2, 3 and 4 to be covered once at Key Stage 1 and twice at Key Stage 2;
- How each year a balance is delivered between Attainment Targets 2, 3 and 4;
- Statements of Attainment to be considered for each term.

strand (i) ask questions, predict and hypothesise;
strand (ii) observe, measure and manipulate variables;
strand (iii) interpret results and evaluate scientific evidence.

The three strands for AT1 will be encouraged in every investigation or activity when appropriate.
Each child will be involved in at least one investigation every term which involves skills from all three strands.

AT 1 Scientific Investigation	Year A						Year B					
	Autumn		Spring		Summer		Autumn		Spring		Summer	
AT 2 Life and living processes												
AT 3 Materials and their properties	SoA		SoA		SoA		SoA		SoA		SoA	
AT 4 Physical processes	SoA										SoA	

149